Insecticides and Pesticides: Techniques for Crop Protection

Insecticides and Pesticides: Techniques for Crop Protection

Edited by
Drew Adams

Larsen & Keller
www.larsen-keller.com

Insecticides and Pesticides: Techniques for Crop Protection
Edited by Drew Adams
ISBN: 978-1-63549-154-8 (Hardback)

© 2017 Larsen & Keller

Larsen & Keller

Published by Larsen and Keller Education,
5 Penn Plaza,
19th Floor,
New York, NY 10001, USA

Cataloging-in-Publication Data

Insecticides and pesticides : techniques for crop protection / edited by Drew Adams.
      p. cm.
Includes bibliographical references and index.
ISBN 978-1-63549-154-8
1. Pesticides. 2. Insecticides. 3. Plants, Protection of. 4. Pests--Control. 5. Plant diseases.
I. Adams, Drew.
SB951 .I67 2017
632.95--dc23

The publisher's policy is to use permanent paper from mills that operate a sustainable forestry policy. Furthermore, the publisher ensures that the text paper and cover boards used have met acceptable environmental accreditation standards.

Printed and bound in the United States of America.

For more information regarding Larsen and Keller Education and its products, please visit the publisher's website www.larsen-keller.com

Table of Contents

Preface

Insecticides and pesticides are chemicals used to protect crops and treat unwanted infestation by insects and pests. They are toxic substances and should be used judiciously. This book attempts to assist those with a goal of delving into the field of insecticides and pesticides treatment and manufacture. For someone with an interest and eye for detail, this book covers the most significant topics in this field. Also included in this textbook is a detailed explanation of the various methods and practices of crop protection. This complex subject is presented in the most comprehensible and easy to understand language. This text aims to serve as a resource guide for students and explain the discipline better.

A short introduction to every chapter is written below to provide an overview of the content of the book:

Chapter 1 - Insecticides and pesticides make crops resistant to insects, their larvae, fungus, birds, and other pests. They help minimize crop loss and increase the yield. Several natural substances act as insecticides and are better alternatives to chemically synthesized insecticides. This introductory chapter on insecticides and pesticides provides in-depth information about the subject matter; **Chapter 2** - There are numerous insecticides in use and in this chapter the reader is introduced to the most widely used ones such as DDT, organophosphate, pyrethroid, permethrin, neonicotinoid, clothianidin, chlorpyrifos etc. These insecticides are a mix of man-made and naturally occurring insecticides and this section provides valuable information on the mode of action, the formulation and sub-types if any; **Chapter 3** - Pesticides can be classified based on the organisms they act on and the toxicity levels they exhibit. Rated on toxicity, they are categorized into toxicity class I, II, III and IV. Based on the agent they are effective against, they can be grouped into fungicide, herbicide, molluscicide and rodenticide. The chapter explores all the categories of pesticides and also acquaints the reader with toxicity labels; **Chapter 4** - Pesticide use is a hotly debated topic due to its potential hazards to the environment, people and other organisms. The information presented in this section draws on these issues and discusses subjects like pesticide resistance, restricted use pesticide, non-pesticide management, pesticide regulation in the United States and pesticide degradation. This chapter elucidates the crucial theories and methods of pesticide use; **Chapter 5** - To ensure that pesticides are delivered to the target organisms efficiently and to minimize the potential risk to people and the environment, the method of pesticide application is as important as the right pesticide. This section explores topics such as aerosol sprays, aerial application and ultra-low volume pesticide application. These techniques improve targeting and decrease spraying inefficiencies; **Chapter 6** - Pests attack plants and damage them decreasing their yield while also affecting soil quality. Integrated pest management consolidates methods and practices used for profitable pest control. This content elaborates on integrated pest management, biopesticides, insect growth regulator, insecticidal soap and fogger. The section on biopesticides sheds light on biologically benign pesticide alternates that are gaining popularity; **Chapter 7** - Pesticides adversely affect the environment and the ecosystem balance. They threaten and sometimes decimate species other than the target organisms. This chapter closely considers the down side to pesticide use by drawing together topics like pollinator decline, health effects of pesticides, pesticide poisoning, environmental impact of pesticides and pesticide drift. This chapter discusses the effects of insecticides and pesticides in a critical manner providing key analysis to the subject matter; **Chapter 8 -** Bees are natural agents responsible for pollination, which in turn allows for genetic variation. The use of pesticides has been linked to a decline in bee populations across the globe. This chapter studies the reasons behind the fall in bee colonies by elaborating on topics such as pesticide toxicity to bees, colony collapse disorder and pesticide formulation; **Chapter 9 -** Plant pathology studies the pathological and environmental conditions that plague plants, disease cycles, plant resistance to disease etc. Plant disease resistance is exhibited either through in-built chemicals and anatomy or immune responses after exposure to diseases. This chapter examines plant pathology, plant disease resistance and pest control.

I extend my sincere thanks to the publisher for considering me worthy of this task. Finally, I thank my family for being a source of support and help.

Editor

Introduction to Pesticides and Insecticides

Insecticides and pesticides make crops resistant to insects, their larvae, fungus, birds, and other pests. They help minimize crop loss and increase the yield. Several natural substances act as insecticides and are better alternatives to chemically synthesized insecticides. This introductory chapter on insecticides and pesticides provides in-depth information about the subject matter.

Pesticide

Pesticides are substances meant for attracting, seducing, and then destroying any pest. They are a class of biocide. The most common use of pesticides is as plant protection products (also known as crop protection products), which in general protect plants from damaging influences such as weeds, fungi, or insects. This use of pesticides is so common that the term *pesticide* is often treated as synonymous with *plant protection product*, although it is in fact a broader term, as pesticides are also used for non-agricultural purposes. The term pesticide includes all of the following: herbicide, insecticide, insect growth regulator, nematicide, termiticide, molluscicide, piscicide, avicide, rodenticide, predacide, bactericide, insect repellent, animal repellent, antimicrobial, fungicide, disinfectant (antimicrobial), and sanitizer.

In general, a pesticide is a chemical or biological agent (such as a virus, bacterium, antimicrobial, or disinfectant) that deters, incapacitates, kills, or otherwise discourages pests. Target pests can include insects, plant pathogens, weeds, mollusks, birds, mammals, fish, nematodes (roundworms), and microbes that destroy property, cause nuisance, or spread disease, or are disease vectors. Although pesticides have benefits, some also have drawbacks, such as potential toxicity to humans and other species. According to the Stockholm Convention on Persistent Organic Pollutants, 9 of the 12 most dangerous and persistent organic chemicals are organochlorine pesticides.

Definition

Type of pesticide	Target pest group
Herbicides	Plant
Algicides or Algaecides	Algae
Avicides	Birds
Bactericides	Bacteria
Fungicides	Fungi and Oomycetes
Insecticides	Insects
Miticides or Acaricides	Mites
Molluscicides	Snails
Nematicides	Nematodes

Rodenticides	Rodents
Virucides	Viruses

The Food and Agriculture Organization (FAO) has defined *pesticide* as:

> any substance or mixture of substances intended for preventing, destroying, or controlling any pest, including vectors of human or animal disease, unwanted species of plants or animals, causing harm during or otherwise interfering with the production, processing, storage, transport, or marketing of food, agricultural commodities, wood and wood products or animal feedstuffs, or substances that may be administered to animals for the control of insects, arachnids, or other pests in or on their bodies. The term includes substances intended for use as a plant growth regulator, defoliant, desiccant, or agent for thinning fruit or preventing the premature fall of fruit. Also used as substances applied to crops either before or after harvest to protect the commodity from deterioration during storage and transport.

Pesticides can be classified by target organism (e.g., herbicides, insecticides, fungicides, rodenticides, and pediculicides - see table), chemical structure (e.g., organic, inorganic, synthetic, or biological (biopesticide), although the distinction can sometimes blur), and physical state (e.g. gaseous (fumigant)). Biopesticides include microbial pesticides and biochemical pesticides. Plant-derived pesticides, or "botanicals", have been developing quickly. These include the pyrethroids, rotenoids, nicotinoids, and a fourth group that includes strychnine and scilliroside.

Many pesticides can be grouped into chemical families. Prominent insecticide families include organochlorines, organophosphates, and carbamates. Organochlorine hydrocarbons (e.g., DDT) could be separated into dichlorodiphenylethanes, cyclodiene compounds, and other related compounds. They operate by disrupting the sodium/potassium balance of the nerve fiber, forcing the nerve to transmit continuously. Their toxicities vary greatly, but they have been phased out because of their persistence and potential to bioaccumulate. Organophosphate and carbamates largely replaced organochlorines. Both operate through inhibiting the enzyme acetylcholinesterase, allowing acetylcholine to transfer nerve impulses indefinitely and causing a variety of symptoms such as weakness or paralysis. Organophosphates are quite toxic to vertebrates, and have in some cases been replaced by less toxic carbamates. Thiocarbamate and dithiocarbamates are subclasses of carbamates. Prominent families of herbicides include phenoxy and benzoic acid herbicides (e.g. 2,4-D), triazines (e.g., atrazine), ureas (e.g., diuron), and Chloroacetanilides (e.g., alachlor). Phenoxy compounds tend to selectively kill broad-leaf weeds rather than grasses. The phenoxy and benzoic acid herbicides function similar to plant growth hormones, and grow cells without normal cell division, crushing the plant's nutrient transport system. Triazines interfere with photosynthesis. Many commonly used pesticides are not included in these families, including glyphosate.

Pesticides can be classified based upon their biological mechanism function or application method. Most pesticides work by poisoning pests. A systemic pesticide moves inside a plant following absorption by the plant. With insecticides and most fungicides, this movement is usually upward (through the xylem) and outward. Increased efficiency may be a result. Systemic insecticides, which poison pollen and nectar in the flowers, may kill bees and other needed pollinators.

In 2009, the development of a new class of fungicides called paldoxins was announced. These work by taking advantage of natural defense chemicals released by plants called phytoalexins, which fungi then detoxify using enzymes. The paldoxins inhibit the fungi's detoxification enzymes. They are believed to be safer and greener.

Uses

Pesticides are used to control organisms that are considered to be harmful. For example, they are used to kill mosquitoes that can transmit potentially deadly diseases like West Nile virus, yellow fever, and malaria. They can also kill bees, wasps or ants that can cause allergic reactions. Insecticides can protect animals from illnesses that can be caused by parasites such as fleas. Pesticides can prevent sickness in humans that could be caused by moldy food or diseased produce. Herbicides can be used to clear roadside weeds, trees and brush. They can also kill invasive weeds that may cause environmental damage. Herbicides are commonly applied in ponds and lakes to control algae and plants such as water grasses that can interfere with activities like swimming and fishing and cause the water to look or smell unpleasant. Uncontrolled pests such as termites and mold can damage structures such as houses. Pesticides are used in grocery stores and food storage facilities to manage rodents and insects that infest food such as grain. Each use of a pesticide carries some associated risk. Proper pesticide use decreases these associated risks to a level deemed acceptable by pesticide regulatory agencies such as the United States Environmental Protection Agency (EPA) and the Pest Management Regulatory Agency (PMRA) of Canada.

DDT, sprayed on the walls of houses, is an organochlorine that has been used to fight malaria since the 1950s. Recent policy statements by the World Health Organization have given stronger support to this approach. However, DDT and other organochlorine pesticides have been banned in most countries worldwide because of their persistence in the environment and human toxicity. DDT use is not always effective, as resistance to DDT was identified in Africa as early as 1955, and by 1972 nineteen species of mosquito worldwide were resistant to DDT.

Amount Used

In 2006 and 2007, the world used approximately 2.4 megatonnes (5.3×10^9 lb) of pesticides, with herbicides constituting the biggest part of the world pesticide use at 40%, followed by insecticides (17%) and fungicides (10%). In 2006 and 2007 the U.S. used approximately 0.5 megatonnes (1.1×10^9 lb) of pesticides, accounting for 22% of the world total, including 857 million pounds (389 kt) of conventional pesticides, which are used in the agricultural sector (80% of conventional pesticide use) as well as the industrial, commercial, governmental and home & garden sectors. Pesticides are also found in majority of U.S. households with 78 million out of the 105.5 million households indicating that they use some form of pesticide. As of 2007, there were more than 1,055 active ingredients registered as pesticides, which yield over 20,000 pesticide products that are marketed in the United States.

The US used some 1 kg (2.2 pounds) per hectare of arable land compared with: 4.7 kg in China, 1.3 kg in the UK, 0.1 kg in Cameroon, 5.9 kg in Japan and 2.5 kg in Italy. Insecticide use in the US has declined by more than half since 1980, (.6%/yr) mostly due to the near phase-out of organophosphates. In corn fields, the decline was even steeper, due to the switchover to transgenic Bt corn.

For the global market of crop protection products, market analysts forecast revenues of over 52 billion US$ in 2019.

Benefits

Pesticides can save farmers' money by preventing crop losses to insects and other pests; in the U.S., farmers get an estimated fourfold return on money they spend on pesticides. One study found that not using pesticides reduced crop yields by about 10%. Another study, conducted in 1999, found that a ban on pesticides in the United States may result in a rise of food prices, loss of jobs, and an increase in world hunger.

There are two levels of benefits for pesticide use, primary and secondary. Primary benefits are direct gains from the use of pesticides and secondary benefits are effects that are more long-term.

Primary Benefits

1. Controlling pests and plant disease vectors

    o   Improved crop/livestock yields

    o   Improved crop/livestock quality

    o   Invasive species controlled

2. Controlling human/livestock disease vectors and nuisance organisms

    o   Human lives saved and suffering reduced

    o   Animal lives saved and suffering reduced

    o   Diseases contained geographically

3. Controlling organisms that harm other human activities and structures

    o   Drivers view unobstructed

    o   Tree/brush/leaf hazards prevented

    o   Wooden structures protected

Monetary

Every dollar ($1) that is spent on pesticides for crops yields four dollars ($4) in crops saved. This means based that, on the amount of money spent per year on pesticides, $10 billion, there is an additional $40 billion savings in crop that would be lost due to damage by insects and weeds. In general, farmers benefit from having an increase in crop yield and from being able to grow a variety of crops throughout the year. Consumers of agricultural products also benefit from being able to afford the vast quantities of produce available year-round. The general public also benefits from the use of pesticides for the control of insect-borne diseases and illnesses, such as malaria. The use of pesticides creates a large job market within the agrichemical sector.

Costs

On the cost side of pesticide use there can be costs to the environment, costs to human health, as well as costs of the development and research of new pesticides.

Health Effects

Pesticides may cause acute and delayed health effects in people who are exposed. Pesticide exposure can cause a variety of adverse health effects, ranging from simple irritation of the skin and eyes to more severe effects such as affecting the nervous system, mimicking hormones causing reproductive problems, and also causing cancer. A 2007 systematic review found that "most studies on non-Hodgkin lymphoma and leukemia showed positive associations with pesticide exposure" and thus concluded that cosmetic use of pesticides should be decreased. There is substantial evidence of associations between organophosphate insecticide exposures and neurobehavioral alterations. Limited evidence also exists for other negative outcomes from pesticide exposure including neurological, birth defects, and fetal death.

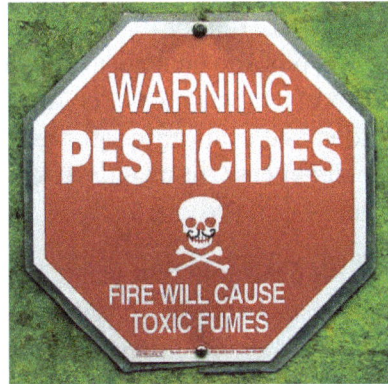

A sign warning about potential pesticide exposure.

The American Academy of Pediatrics recommends limiting exposure of children to pesticides and using safer alternatives:

The World Health Organization and the UN Environment Programme estimate that each year, 3 million workers in agriculture in the developing world experience severe poisoning from pesticides, about 18,000 of whom die. Owing to inadequate regulation and safety precautions, 99% of pesticide related deaths occur in developing countries that account for only 25% of pesticide usage. According to one study, as many as 25 million workers in developing countries may suffer mild pesticide poisoning yearly. There are several careers aside from agriculture that may also put individuals at risk of health effects from pesticide exposure including pet groomers, groundskeepers, and fumigators.

One study found pesticide self-poisoning the method of choice in one third of suicides worldwide, and recommended, among other things, more restrictions on the types of pesticides that are most harmful to humans.

A 2014 epidemiological review found associations between autism and exposure to certain pesticides, but noted that the available evidence was insufficient to conclude that the relationship was causal.

Environmental Effect

Pesticide use raises a number of environmental concerns. Over 98% of sprayed insecticides and 95% of herbicides reach a destination other than their target species, including non-target species, air, water and soil. Pesticide drift occurs when pesticides suspended in the air as particles are carried by wind to other areas, potentially contaminating them. Pesticides are one of the causes of water pollution, and some pesticides are persistent organic pollutants and contribute to soil contamination.

In addition, pesticide use reduces biodiversity, contributes to pollinator decline, destroys habitat (especially for birds), and threatens endangered species. Pests can develop a resistance to the pesticide (pesticide resistance), necessitating a new pesticide. Alternatively a greater dose of the pesticide can be used to counteract the resistance, although this will cause a worsening of the ambient pollution problem.

Since chlorinated hydrocarbon pesticides dissolve in fats and are not excreted, organisms tend to retain them almost indefinitely. Biological magnification is the process whereby these chlorinated hydrocarbons (pesticides) are more concentrated at each level of the food chain. Among marine animals, pesticide concentrations are higher in carnivorous fishes, and even more so in the fish-eating birds and mammals at the top of the ecological pyramid. Global distillation is the process whereby pesticides are transported from warmer to colder regions of the Earth, in particular the Poles and mountain tops. Pesticides that evaporate into the atmosphere at relatively high temperature can be carried considerable distances (thousands of kilometers) by the wind to an area of lower temperature, where they condense and are carried back to the ground in rain or snow.

In order to reduce negative impacts, it is desirable that pesticides be degradable or at least quickly deactivated in the environment. Such loss of activity or toxicity of pesticides is due to both innate chemical properties of the compounds and environmental processes or conditions. For example, the presence of halogens within a chemical structure often slows down degradation in an aerobic environment. Adsorption to soil may retard pesticide movement, but also may reduce bioavailability to microbial degraders.

Economics

Human health and environmental cost from pesticides in the United States is estimated at $9.6 billion offset by about $40 billion in increased agricultural production:

Harm	Annual US cost
Public health	$1.1 billion
Pesticide resistance in pest	$1.5 billion
Crop losses caused by pesticides	$1.4 billion
Bird losses due to pesticides	$2.2 billion
Groundwater contamination	$2.0 billion
Other costs	$1.4 billion
Total costs	**$9.6 billion**

Additional costs include the registration process and the cost of purchasing pesticides. The registration process can take several years to complete (there are 70 different types of field test) and can cost \$50–70 million for a single pesticide. Annually the United States spends \$10 billion on pesticides.

Alternatives

Alternatives to pesticides are available and include methods of cultivation, use of biological pest controls (such as pheromones and microbial pesticides), genetic engineering, and methods of interfering with insect breeding. Application of composted yard waste has also been used as a way of controlling pests. These methods are becoming increasingly popular and often are safer than traditional chemical pesticides. In addition, EPA is registering reduced-risk conventional pesticides in increasing numbers.

Cultivation practices include polyculture (growing multiple types of plants), crop rotation, planting crops in areas where the pests that damage them do not live, timing planting according to when pests will be least problematic, and use of trap crops that attract pests away from the real crop. In the U.S., farmers have had success controlling insects by spraying with hot water at a cost that is about the same as pesticide spraying.

Release of other organisms that fight the pest is another example of an alternative to pesticide use. These organisms can include natural predators or parasites of the pests. Biological pesticides based on entomopathogenic fungi, bacteria and viruses cause disease in the pest species can also be used.

Interfering with insects' reproduction can be accomplished by sterilizing males of the target species and releasing them, so that they mate with females but do not produce offspring. This technique was first used on the screwworm fly in 1958 and has since been used with the medfly, the tsetse fly, and the gypsy moth. However, this can be a costly, time consuming approach that only works on some types of insects.

Agroecology emphasize nutrient recycling, use of locally available and renewable resources, adaptation to local conditions, utilization of microenvironments, reliance on indigenous knowledge and yield maximization while maintaining soil productivity. Agroecology also emphasizes empowering people and local communities to contribute to development, and encouraging "multi-directional" communications rather than the conventional "top-down" method.

Push Pull Strategy

The term "push-pull" was established in 1987 as an approach for integrated pest management (IPM). This strategy uses a mixture of behavior-modifying stimuli to manipulate the distribution and abundance of insects. "Push" means the insects are repelled or deterred away from whatever resource that is being protected. "Pull" means that certain stimuli (semiochemical stimuli, pheromones, food additives, visual stimuli, genetically altered plants, etc.) are used to attract pests to trap crops where they will be killed. There are numerous different components involved in order to implement a Push-Pull Strategy in IPM.

Many case studies testing the effectiveness of the push-pull approach have been done across the world. The most successful push-pull strategy was developed in Africa for subsistence farming. Another successful case study was performed on the control of *Helicoverpa* in cotton crops in Australia. In Europe, the Middle East, and the United States, push-pull strategies were successfully used in the controlling of *Sitona lineatus* in bean fields.

Some advantages of using the push-pull method are less use of chemical or biological materials and better protection against insect habituation to this control method. Some disadvantages of the push-pull strategy is that if there is a lack of appropriate knowledge of behavioral and chemical ecology of the host-pest interactions then this method becomes unreliable. Furthermore, because the push-pull method is not a very popular method of IPM operational and registration costs are higher.

Effectiveness

Some evidence shows that alternatives to pesticides can be equally effective as the use of chemicals. For example, Sweden has halved its use of pesticides with hardly any reduction in crops. In Indonesia, farmers have reduced pesticide use on rice fields by 65% and experienced a 15% crop increase. A study of Maize fields in northern Florida found that the application of composted yard waste with high carbon to nitrogen ratio to agricultural fields was highly effective at reducing the population of plant-parasitic nematodes and increasing crop yield, with yield increases ranging from 10% to 212%; the observed effects were long-term, often not appearing until the third season of the study.

However, pesticide resistance is increasing. In the 1940s, U.S. farmers lost only 7% of their crops to pests. Since the 1980s, loss has increased to 13%, even though more pesticides are being used. Between 500 and 1,000 insect and weed species have developed pesticide resistance since 1945.

Types

Pesticides are often referred to according to the type of pest they control. Pesticides can also be considered as either biodegradable pesticides, which will be broken down by microbes and other living beings into harmless compounds, or persistent pesticides, which may take months or years before they are broken down: it was the persistence of DDT, for example, which led to its accumulation in the food chain and its killing of birds of prey at the top of the food chain. Another way to think about pesticides is to consider those that are chemical pesticides or are derived from a common source or production method.

Some examples of chemically-related pesticides are:

Organophosphate Pesticides

Organophosphates affect the nervous system by disrupting, acetylcholinesterase activity, the enzyme that regulates acetylcholine, a neurotransmitter. Most organophosphates are insecticides. They were developed during the early 19th century, but their effects on insects, which are similar to their effects on humans, were discovered in 1932. Some are very poisonous. However, they usually are not persistent in the environment.

Carbamate Pesticides

Carbamate pesticides affect the nervous system by disrupting an enzyme that regulates acetylcholine, a neurotransmitter. The enzyme effects are usually reversible. There are several subgroups within the carbamates.

Organochlorine Insecticides

They were commonly used in the past, but many have been removed from the market due to their health and environmental effects and their persistence (e.g., DDT, chlordane, and toxaphene).

Pyrethroid Pesticides

They were developed as a synthetic version of the naturally occurring pesticide pyrethrin, which is found in chrysanthemums. They have been modified to increase their stability in the environment. Some synthetic pyrethroids are toxic to the nervous system.

Sulfonylurea Herbicides

The following sulfonylureas have been commercialized for weed control: amidosulfuron, azimsulfuron, bensulfuron-methyl, chlorimuron-ethyl, ethoxysulfuron, flazasulfuron, flupyrsulfuron-methyl-sodium, halosulfuron-methyl, imazosulfuron, nicosulfuron, oxasulfuron, primisulfuron-methyl, pyrazosulfuron-ethyl, rimsulfuron, sulfometuron-methyl Sulfosulfuron, terbacil, bispyribac-sodium, cyclosulfamuron, and pyrithiobac-sodium. Nicosulfuron, triflusulfuron methyl, and chlorsulfuron are broad-spectrum herbicides that kill plants by inhibiting the enzyme acetolactate synthase. In the 1960s, more than 1 kg/ha (0.89 lb/acre) crop protection chemical was typically applied, while sulfonylureates allow as little as 1% as much material to achieve the same effect.

Biopesticides

Biopesticides are certain types of pesticides derived from such natural materials as animals, plants, bacteria, and certain minerals. For example, canola oil and baking soda have pesticidal applications and are considered biopesticides. Biopesticides fall into three major classes:

- Microbial pesticides which consist of bacteria, entomopathogenic fungi or viruses (and sometimes includes the metabolites that bacteria or fungi produce). Entomopathogenic nematodes are also often classed as microbial pesticides, even though they are multi-cellular.

- Biochemical pesticides or herbal pesticides are naturally occurring substances that control (or monitor in the case of pheromones) pests and microbial diseases.

- Plant-incorporated protectants (PIPs) have genetic material from other species incorporated into their genetic material (*i.e.* GM crops). Their use is controversial, especially in many European countries.

Classified by Type of Pest

Pesticides that are related to the type of pests are:

Type	Action
Algicides	Control algae in lakes, canals, swimming pools, water tanks, and other sites
Antifouling agents	Kill or repel organisms that attach to underwater surfaces, such as boat bottoms

Antimicrobials	Kill microorganisms (such as bacteria and viruses)
Attractants	Attract pests (for example, to lure an insect or rodent to a trap). (However, food is not considered a pesticide when used as an attractant.)
Biopesticides	Biopesticides are certain types of pesticides derived from such natural materials as animals, plants, bacteria, and certain minerals
Biocides	Kill microorganisms
Disinfectants and sanitizers	Kill or inactivate disease-producing microorganisms on inanimate objects
Fungicides	Kill fungi (including blights, mildews, molds, and rusts)
Fumigants	Produce gas or vapor intended to destroy pests in buildings or soil
Herbicides	Kill weeds and other plants that grow where they are not wanted
Insecticides	Kill insects and other arthropods
Miticides	Kill mites that feed on plants and animals
Microbial pesticides	Microorganisms that kill, inhibit, or out compete pests, including insects or other microorganisms
Molluscicides	Kill snails and slugs
Nematicides	Kill nematodes (microscopic, worm-like organisms that feed on plant roots)
Ovicides	Kill eggs of insects and mites
Pheromones	Biochemicals used to disrupt the mating behavior of insects
Repellents	Repel pests, including insects (such as mosquitoes) and birds
Rodenticides	Control mice and other rodents

Further Types of Pesticides

The term pesticide also include these substances:

Defoliants : Cause leaves or other foliage to drop from a plant, usually to facilitate harvest. Desiccants : Promote drying of living tissues, such as unwanted plant tops. Insect growth regulators : Disrupt the molting, maturity from pupal stage to adult, or other life processes of insects. Plant growth regulators : Substances (excluding fertilizers or other plant nutrients) that alter the expected growth, flowering, or reproduction rate of plants.

Regulation

International

In most countries, pesticides must be approved for sale and use by a government agency.

In Europe, recent EU legislation has been approved banning the use of highly toxic pesticides including those that are carcinogenic, mutagenic or toxic to reproduction, those that are endocrine-disrupting, and those that are persistent, bioaccumulative and toxic (PBT) or very persistent and very bioaccumulative (vPvB). Measures were approved to improve the general safety of pesticides across all EU member states.

Though pesticide regulations differ from country to country, pesticides, and products on which they were used are traded across international borders. To deal with inconsistencies in regulations among countries, delegates to a conference of the United Nations Food and Agriculture Organization adopted an International Code of Conduct on the Distribution and Use of Pesticides in 1985 to create voluntary standards of pesticide regulation for different countries. The Code was updated in 1998 and 2002. The FAO claims that the code has raised awareness about pesticide hazards and decreased the number of countries without restrictions on pesticide use.

Three other efforts to improve regulation of international pesticide trade are the United Nations London Guidelines for the Exchange of Information on Chemicals in International Trade and the United Nations Codex Alimentarius Commission. The former seeks to implement procedures for ensuring that prior informed consent exists between countries buying and selling pesticides, while the latter seeks to create uniform standards for maximum levels of pesticide residues among participating countries. Both initiatives operate on a voluntary basis.

Pesticides safety education and pesticide applicator regulation are designed to protect the public from pesticide misuse, but do not eliminate all misuse. Reducing the use of pesticides and choosing less toxic pesticides may reduce risks placed on society and the environment from pesticide use. Integrated pest management, the use of multiple approaches to control pests, is becoming widespread and has been used with success in countries such as Indonesia, China, Bangladesh, the U.S., Australia, and Mexico. IPM attempts to recognize the more widespread impacts of an action on an ecosystem, so that natural balances are not upset. New pesticides are being developed, including biological and botanical derivatives and alternatives that are thought to reduce health and environmental risks. In addition, applicators are being encouraged to consider alternative controls and adopt methods that reduce the use of chemical pesticides.

Pesticides can be created that are targeted to a specific pest's lifecycle, which can be environmentally more friendly. For example, potato cyst nematodes emerge from their protective cysts in response to a chemical excreted by potatoes; they feed on the potatoes and damage the crop. A similar chemical can be applied to fields early, before the potatoes are planted, causing the nematodes to emerge early and starve in the absence of potatoes.

United States

Preparation for an application of hazardous herbicide in USA.

In the United States, the Environmental Protection Agency (EPA) is responsible for regulating pesticides under the Federal Insecticide, Fungicide, and Rodenticide Act (FIFRA) and the Food Quality Protection Act (FQPA). Studies must be conducted to establish the conditions in which the material is safe to use and the effectiveness against the intended pest(s). The EPA regulates pesticides to ensure that these products do not pose adverse effects to humans or the environment. Pesticides produced before November 1984 continue to be reassessed in order to meet the current scientific and regulatory standards. All registered pesticides are reviewed every 15 years to ensure they meet the proper standards. During the registration process, a label is created. The label contains directions for proper use of the material in addition to safety restrictions. Based on acute toxicity, pesticides are assigned to a Toxicity Class.

Some pesticides are considered too hazardous for sale to the general public and are designated restricted use pesticides. Only certified applicators, who have passed an exam, may purchase or supervise the application of restricted use pesticides. Records of sales and use are required to be maintained and may be audited by government agencies charged with the enforcement of pesticide regulations. These records must be made available to employees and state or territorial environmental regulatory agencies.

The EPA regulates pesticides under two main acts, both of which amended by the Food Quality Protection Act of 1996. In addition to the EPA, the United States Department of Agriculture (USDA) and the United States Food and Drug Administration (FDA) set standards for the level of pesticide residue that is allowed on or in crops. The EPA looks at what the potential human health and environmental effects might be associated with the use of the pesticide.

In addition, the U.S. EPA uses the National Research Council's four-step process for human health risk assessment: (1) Hazard Identification, (2) Dose-Response Assessment, (3) Exposure Assessment, and (4) Risk Characterization.

Recently Kaua'i County (Hawai'i) passed Bill No. 2491 to add an article to Chapter 22 of the county's code relating to pesticides and GMOs. The bill strengthens protections of local communities in Kaua'i where many large pesticide companies test their products.

History

Since before 2000 BC, humans have utilized pesticides to protect their crops. The first known pesticide was elemental sulfur dusting used in ancient Sumer about 4,500 years ago in ancient Mesopotamia. The Rig Veda, which is about 4,000 years old, mentions the use of poisonous plants for pest control. By the 15th century, toxic chemicals such as arsenic, mercury, and lead were being applied to crops to kill pests. In the 17th century, nicotine sulfate was extracted from tobacco leaves for use as an insecticide. The 19th century saw the introduction of two more natural pesticides, pyrethrum, which is derived from chrysanthemums, and rotenone, which is derived from the roots of tropical vegetables. Until the 1950s, arsenic-based pesticides were dominant. Paul Müller discovered that DDT was a very effective insecticide. Organochlorines such as DDT were dominant, but they were replaced in the U.S. by organophosphates and carbamates by 1975. Since then, pyrethrin compounds have become the dominant insecticide. Herbicides became common in the 1960s, led by "triazine and other nitrogen-based compounds, carboxylic acids such as 2,4-dichlorophenoxyacetic acid, and glyphosate".

The first legislation providing federal authority for regulating pesticides was enacted in 1910; however, decades later during the 1940s manufacturers began to produce large amounts of synthetic pesticides and their use became widespread. Some sources consider the 1940s and 1950s to have been the start of the "pesticide era." Although the U.S. Environmental Protection Agency was established in 1970 and amendments to the pesticide law in 1972, pesticide use has increased 50-fold since 1950 and 2.3 million tonnes (2.5 million short tons) of industrial pesticides are now used each year. Seventy-five percent of all pesticides in the world are used in developed countries, but use in developing countries is increasing. A study of USA pesticide use trends through 1997 was published in 2003 by the National Science Foundation's Center for Integrated Pest Management.

In the 1960s, it was discovered that DDT was preventing many fish-eating birds from reproducing, which was a serious threat to biodiversity. Rachel Carson wrote the best-selling book *Silent Spring* about biological magnification. The agricultural use of DDT is now banned under the Stockholm Convention on Persistent Organic Pollutants, but it is still used in some developing nations to prevent malaria and other tropical diseases by spraying on interior walls to kill or repel mosquitoes.

Insecticide

An insecticide is a substance used to kill insects. They include ovicides and larvicides used against insect eggs and larvae, respectively. Insecticides are used in agriculture, medicine, industry and by consumers. Insecticides are claimed to be a major factor behind the increase in agricultural 20th century's productivity. Nearly all insecticides have the potential to significantly alter ecosystems; many are toxic to humans; some concentrate along the food chain.

Insecticides can be classified in two major groups: systemic insecticides, which have residual or long term activity; and contact insecticides, which have no residual activity.

Furthermore, one can distinguish three types of insecticide. 1. Natural insecticides, such as nicotine, pyrethrum and neem extracts, made by plants as defenses against insects. 2. Inorganic insecticides, which are metals. 3. Organic insecticides, which are organic chemical compounds, mostly working by contact.

The mode of action describes how the pesticide kills or inactivates a pest. It provides another way of classifying insecticides. Mode of action is important in understanding whether an insecticide will be toxic to unrelated species, such as fish, birds and mammals.

Insecticides are distinct from insect repellents, which do not kill.

Type of Activity

Systemic insecticides become incorporated and distributed systemically throughout the whole plant. When insects feed on the plant, they ingest the insecticide. Systemic insecticides produced by transgenic plants are called plant-incorporated protectants (PIPs). For instance, a gene that codes for a specific Bacillus thuringiensis biocidal protein was introduced into corn and other species. The plant manufactures the protein, which kills the insect when consumed. Systemic insecticides have activity pertaining to their residue which is called "residual activity" or long-term activity.

Contact insecticides are toxic to insects upon direct contact. These can be inorganic insecticides, which are metals and include arsenates, copper and fluorine compounds, which are less commonly used, and the commonly used sulfur. Contact insecticides can be organic insecticides, i.e. organic chemical compounds, synthetically produced, and comprising the largest numbers of pesticides used today. Or they can be natural compounds like pyrethrum, neem oil etc. Contact insecticides usually have no residual activity.

Efficacy can be related to the quality of pesticide application, with small droplets, such as aerosols often improving performance.

Major Classes

Organochlorides

The best known organochloride, DDT, was created by Swiss scientist Paul Müller. For this discovery, he was awarded the 1948 Nobel Prize for Physiology or Medicine. DDT was introduced in 1944. It functions by opening sodium channels in the insect's nerve cells. The contemporaneous rise of the chemical industry facilitated large-scale production of DDT and related chlorinated hydrocarbons.

Organophosphates and Carbamates

Organophosphates are another large class of contact insecticides. These also target the insect's nervous system. Organophosphates interfere with the enzymes acetylcholinesterase and other cholinesterases, disrupting nerve impulses and killing or disabling the insect. Organophosphate insecticides and chemical warfare nerve agents (such as sarin, tabun, soman, and VX) work in the same way. Organophosphates have a cumulative toxic effect to wildlife, so multiple exposures to the chemicals amplifies the toxicity. In the US, organophosphate use declined with the rise of substitutes.

Carbamate insecticides have similar mechanisms to organophosphates, but have a much shorter duration of action and are somewhat less toxic.

Pyrethroids

Pyrethroid pesticides mimic the insecticidal activity of the natural compound pyrethrum, which is found in pyrethrins. These compounds are nonpersistent sodium channel modulators and are less toxic than organophosphates and carbamates. Compounds in this group are often applied against household pests.

Neonicotinoids

Neonicotinoids are synthetic analogues of the natural insecticide nicotine (with much lower acute mammalian toxicity and greater field persistence). These chemicals are acetylcholine receptor agonists. They are broad-spectrum systemic insecticides, with rapid action (minutes-hours). They are applied as sprays, drenches, seed and soil treatments. Treated insects exhibit leg tremors, rapid wing motion, stylet withdrawal (aphids), disoriented movement, paralysis and death. Imidacloprid may be the most common. It has recently come under scrutiny for allegedly pernicious effects on honeybees and its potential to increase the susceptibility of rice to planthopper attacks.

Ryanoids

Ryanoids are synthetic analogues with the same mode of action as ryanodine, a naturally occurring insecticide extracted from *Ryania speciosa* (Flacourtiaceae). They bind to calcium channels in cardiac and skeletal muscle, blocking nerve transmission. Only one such insecticide is currently registered, Rynaxypyr, generic name chlorantraniliprole.

Plant-incorporated Protectants

Transgenic crops that act as insecticides began in 1996 with a genetically modified potato that produced the Cry protein, derived from the bacterium Bacillus thuringiensis, which is toxic to beetle larvae such as the Colorado potato beetle. The technique has been expanded to include the use of RNA interference RNAi that fatally silences crucial insect genes. RNAi likely evolved as a defense against viruses. Midgut cells in many larvae take up the molecules and help spread the signal. The technology can target only insects that have the silenced sequence, as was demonstrated when a particular RNAi affected only one of four fruit fly species. The technique is expected to replace many other insecticides, which are losing effectiveness due to the spread of pesticide resistance.

Biological

Many plants exude substances to repel insects. Premier examples are substances activated by the enzyme myrosinase. This enzyme converts glucosinolates to various compounds that are toxic to herbivorous insects. One product of this enzyme is allyl isothiocyanate, the pungent ingredient in horseradish sauces.

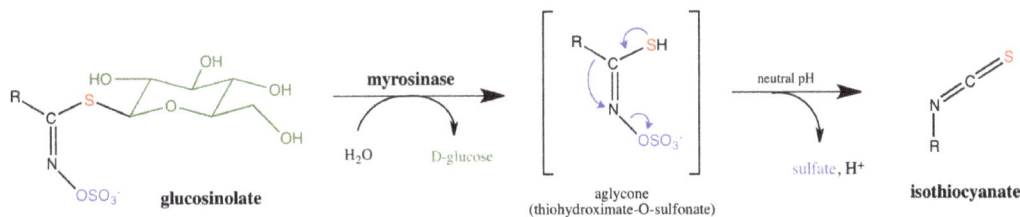

Biosynthesis of Antifeedants by The Action of Myrosinase.

The myrosinase is released only upon crushing the flesh of horseradish. Since allyl isothiocyanate is harmful to the plant as well as the insect, it is stored in the harmless form of the glucosinolate, separate from the myrosinase enzyme.

In general, tree rosin is considered a natural insecticide. To be specific, the production of oleoresin by conifer species is a component of the defense response against insect attack and fungal pathogen infection.

Bacterial

Bacillus thuringiensis is a bacterial disease that affects Lepidopterans and some other insects. Toxins produced by strains of this bacterium are used as a larvicide against caterpillars, beetles, and mosquitoes. Toxins from *Saccharopolyspora spinosa* are isolated from fermentations and sold as

Spinosad. Because these toxins have little effect on other organisms, they are considered more environmentally friendly than synthetic pesticides. The toxin from *B. thuringiensis* (Bt toxin) has been incorporated directly into plants through the use of genetic engineering. Other biological insecticides include products based on entomopathogenic fungi (e.g., *Beauveria bassiana*, *Metarhizium anisopliae*), nematodes (e.g., *Steinernema feltiae*) and viruses (e.g., *Cydia pomonella* granulovirus).

Insect Growth Regulators

Insect growth regulator (IGR) is a term coined to include insect hormone mimics and an earlier class of chemicals, the benzoylphenyl ureas, which inhibit chitin(exoskeleton) biosynthesis in insects. Diflubenzuron is a member of the latter class, used primarily to control caterpillars that are pests. The most successful insecticides in this class are the juvenoids (juvenile hormone analogues). Of these, methoprene is most widely used. It has no observable acute toxicity in rats and is approved by World Health Organization (WHO) for use in drinking water cisterns to combat malaria. Most of its uses are to combat insects where the adult is the pest, including mosquitoes, several fly species, and fleas. Two very similar products, hydroprene and kinoprene, are used for controlling species such as cockroaches and white flies. Methoprene was registered with the EPA in 1975. Virtually no reports of resistance have been filed. A more recent type of IGR is the ecdysone agonist tebufenozide (MIMIC), which is used in forestry and other applications for control of caterpillars, which are far more sensitive to its hormonal effects than other insect orders.

Environmental Effects

Effects on Nontarget Species

Some insecticides kill or harm other creatures in addition to those they are intended to kill. For example, birds may be poisoned when they eat food that was recently sprayed with insecticides or when they mistake an insecticide granule on the ground for food and eat it.

Sprayed insecticide may drift from the area to which it is applied and into wildlife areas, especially when it is sprayed aerially.

DDT

The development of DDT was motivated by desire to replace more dangerous or less effective alternatives. DDT was introduced to replace lead and arsenic-based compounds, which were in widespread use in the early 1940s.

DDT was brought to public attention by Rachel Carson's book *Silent Spring*. One side-effect of DDT is to reduce the thickness of shells on the eggs of predatory birds. The shells sometimes become too thin to be viable, reducing bird populations. This occurs with DDT and related compounds due to the process of bioaccumulation, wherein the chemical, due to its stability and fat solubility, accumulates in organisms' fatty tissues. Also, DDT may biomagnify, which causes progressively higher concentrations in the body fat of animals farther up the food chain. The near-worldwide ban on agricultural use of DDT and related chemicals has allowed some of these birds, such as the peregrine falcon, to recover in recent years. A number of organochlorine pesticides have been banned from most uses worldwide. Globally they are controlled via the Stockholm Convention on persistent organic pollutants. These include: aldrin, chlordane, DDT, dieldrin, endrin, heptachlor, mirex and toxaphene.

Pollinator Decline

Insecticides can kill bees and may be a cause of pollinator decline, the loss of bees that pollinate plants, and colony collapse disorder (CCD), in which worker bees from a beehive or Western honey bee colony abruptly disappear. Loss of pollinators means a reduction in crop yields. Sublethal doses of insecticides (i.e. imidacloprid and other neonicotinoids) affect bee foraging behavior. However, research into the causes of CCD was inconclusive as of June 2007.

Examples

Organochlorides	Pyrethroids
• Aldrin	• Allethrin
• Chlordane	• Bifenthrin
• Chlordecone	• Cyhalothrin, Lambda-cyhalothrin
• DDT	• Cypermethrin
• Dieldrin	• Cyfluthrin
• Endosulfan	• Deltamethrin
• Endrin	• Etofenprox
• Heptachlor	• Fenvalerate
• Hexachlorobenzene	• Permethrin
• Lindane (gamma-hexachlorocyclohexane)	• Phenothrin
• Methoxychlor	• Prallethrin
• Mirex	• Resmethrin
• Pentachlorophenol	• Tetramethrin
• TDE	• Tralomethrin
Organophosphates	• Transfluthrin
• Acephate	**Neonicotinoids**
• Azinphos-methyl	• Acetamiprid
• Bensulide	• Clothianidin
• Chlorethoxyfos	• Imidacloprid
• Chlorpyrifos	• Nithiazine
• Chlorpyriphos-methyl	• Thiacloprid
• Diazinon	• Thiamethoxam
• Dichlorvos (DDVP)	**Ryanoids**
• Dicrotophos	• Chlorantraniliprole
• Dimethoate	• Cyantraniliprole
• Disulfoton	• Flubendiamide
• Ethoprop	**Insect growth regulators**
• Fenamiphos	• Benzoylureas
• Fenitrothion	o Diflubenzuron
• Fenthion	o Flufenoxuron
• Fosthiazate	• Cyromazine
• Malathion	• Methoprene
• Methamidophos	• Hydroprene

• Methidathion	• Tebufenozide
• Mevinphos	**Plant-derived**
• Monocrotophos	• Anabasine
• Naled	• Anethole (mosquito larvae)
• Omethoate	• Annonin
• Oxydemeton-methyl	• Asimina (pawpaw tree seeds) for lice
• Parathion	• Azadirachtin
• Parathion-methyl	• Caffeine
• Phorate	• Carapa
• Phosalone	• Cinnamaldehyde (very effective for killing mosquito larvae)
• Phosmet	• Cinnamon leaf oil (very effective for killing mosquito larvae)
• Phostebupirim	• Cinnamyl acetate (kills mosquito larvae)
• Phoxim	• Citral
• Pirimiphos-methyl	• Deguelin
• Profenofos	• Derris
• Terbufos	• Derris (rotenone)
• Tetrachlorvinphos	• *Desmodium caudatum* (leaves and roots)
• Tribufos	• Eugenol (mosquito larvae)
• Trichlorfon	• Linalool
Carbamates	• Myristicin
• Aldicarb	• Neem (Azadirachtin)
• Bendiocarb	• *Nicotiana rustica* (nicotine)
• Carbofuran	• *Peganum harmala*, seeds (smoke from), root
• Carbaryl	• Oregano oil kills beetles *Rhizopertha dominica* (bug found in stored cereal)
• Dioxacarb	
Fenobucarb	Polyketide
• Fenoxycarb	• Pyrethrum
• Isoprocarb	• *Quassia* (South American plant genus)
• Methomyl	• Ryanodine
• 2-(1-Methylpropyl)phenyl methylcarbamate	• Tetranortriterpenoid
	• Thymol (controls varroa mites in bee colonies)
	Biologicals
	• Bacillus sphaericus
	• Bacillus thuringiensis
	• Bacillus thuringiensis aizawi
	• Bacillus thuringiensis israelensis
	• Bacillus thuringiensis kurstaki
	• Bacillus thuringiensis tenebrionis
	• Nuclear Polyhedrosis virus
	• Granulovirus

	• Spinosad AKA Spinosyn A
	• Spinosyn D
	• Lecanicillium lecanii
	Other
	• Diatomaceous earth
	• Borate
	• Borax
	• Boric Acid

References

- Food and Agriculture Organization of the United Nations (2002), International Code of Conduct on the Distribution and Use of Pesticides. Retrieved on 2007-10-25.

- Cornell University. Toxicity of pesticides. Pesticide fact sheets and tutorial, module 4. Pesticide Safety Education Program. Retrieved on 2007-10-10.

- Helfrich, LA, Weigmann, DL, Hipkins, P, and Stinson, ER (June 1996), Pesticides and aquatic animals: A guide to reducing impacts on aquatic systems. Virginia Cooperative Extension. Retrieved on 2007-10-14.

- World Health Organization (September 15, 2006), WHO gives indoor use of DDT a clean bill of health for controlling malaria. Retrieved on September 13, 2007.

- "CDC - Pesticide Illness & Injury Surveillance - NIOSH Workplace Safety and Health Topic". Cdc.gov. 2013-09-11. Retrieved 2014-01-28.

- Kellogg RL, Nehring R, Grube A, Goss DW, and Plotkin S (February 2000), Environmental indicators of pesticide leaching and runoff from farm fields. United States Department of Agriculture Natural Resources Conservation Service. Retrieved on 2007-10-03.

- Pimentel, David, H. Acquay, M. Biltonen, P. Rice, and M. Silva. "Environmental and Economic Costs of Pesticide Use." BioScience 42.10 (1992): 750-60., . Retrieved on February 25, 2011.

- "CDC - Pesticide Illness & Injury Surveillance - NIOSH Workplace Safety and Health Topic". www.cdc.gov. Retrieved 2016-02-11.

- Wells, M (March 11, 2007). "Vanishing bees threaten U.S. crops". www.bbc.co.uk. London: BBC News. Retrieved 2007-09-19.

- Palmer, WE, Bromley, PT, and Brandenburg, RL. Wildlife & pesticides - Peanuts. North Carolina Cooperative Extension Service. Retrieved on 2007-10-11.

- Pimentel, David. "Environmental and Economic Costs of the Application of Pesticides Primarily in the United States" Environment, Development and Sustainability 7 (2005): 229-252. Retrieved on February 25, 2011.

- SP-401 Skylab, Classroom in Space: Part III - Science Demonstrations, Chapter 17: Life Sciences. History.nasa.gov. Retrieved on September 17, 2007.

- "Pesticides 101 - A primer on pesticides, their use in agriculture and the exposure we face | Pesticide Action Network". Panna.org. Retrieved 2014-01-28.

- Francis Borgio J, Sahayaraj K and Alper Susurluk I (eds) . Microbial Insecticides: Principles and Applications, Nova Publishers, USA. 492pp. ISBN 978-1-61209-223-2

- Food and Agriculture Organization of the United Nations, Programmes: International Code of Conduct on the Distribution and Use of Pesticides. Retrieved on 2007-10-25.

- Science Daily, (October 11, 2001), Environmentally-friendly pesticide to combat potato cyst nematodes. Sciencedaily.com. Retrieved on September 19, 2007.

Types of Insecticides

There are numerous insecticides in use and in this chapter the reader is introduced to the most widely used ones such as DDT, organophosphate, pyrethroid, permethrin, neonicotinoid, clothianidin, chlorpyrifos etc. These insecticides are a mix of man-made and naturally occurring insecticides and this section provides valuable information on the mode of action, the formulation and sub-types if any.

DDT

DDT (dichlorodiphenyltrichloroethane) is a colorless, crystalline, tasteless and almost odorless organochlorine known for its insecticidal properties and environmental impacts. DDT has been formulated in multiple forms, including solutions in xylene or petroleum distillates, emulsifiable concentrates, water-wettable powders, granules, aerosols, smoke candles and charges for vaporizers and lotions.

First synthesized in 1874, DDT's insecticidal action was discovered by the Swiss chemist Paul Hermann Müller in 1939. It was used in the second half of World War II to control malaria and typhus among civilians and troops. After the war, DDT was also used as an agricultural insecticide and its production and use duly increased. Müller was awarded the Nobel Prize in Physiology or Medicine "for his discovery of the high efficiency of DDT as a contact poison against several arthropods" in 1948.

In 1962, Rachel Carson's book *Silent Spring* was published. It cataloged the environmental impacts of widespread DDT spraying in the United States and questioned the logic of releasing large amounts of potentially dangerous chemicals into the environment without understanding their effects on the environment or human health. The book claimed that DDT and other pesticides had been shown to cause cancer and that their agricultural use was a threat to wildlife, particularly birds. Its publication was a seminal event for the environmental movement and resulted in a large public outcry that eventually led, in 1972, to a ban on DDT's agricultural use in the United States. A worldwide ban on agricultural use was formalized under the Stockholm Convention on Persistent Organic Pollutants, but its limited and still-controversial use in disease vector control continues, because of its effectiveness in reducing malarial infections, balanced by environmental and other health concerns.

Along with the passage of the Endangered Species Act, the United States ban on DDT is cited by scientists as a major factor in the comeback of the bald eagle (the national bird of the United States) and the peregrine falcon from near-extinction in the contiguous United States.

Properties and Chemistry

DDT is similar in structure to the insecticide methoxychlor and the acaricide dicofol. It is highly hydrophobic and nearly insoluble in water but has good solubility in most organic solvents, fats and oils. DDT does not occur naturally. It is produced by the reaction of chloral (CCl_3CHO) with chlorobenzene (C_6H_5Cl) in the presence of a sulfuric acid catalyst. DDT has been marketed under trade names including Anofex, Cezarex, Chlorophenothane, Clofenotane, Dicophane, Dinocide, Gesarol, Guesapon, Guesarol, Gyron, Ixodex, Neocid, Neocidol and Zerdane.

Isomers and Related Compounds

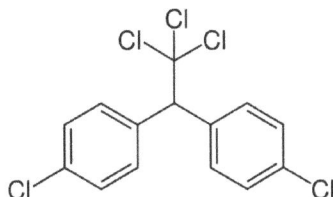

o,p'-DDT, a minor component in commercial DDT.

Commercial DDT is a mixture of several closely–related compounds. The major component (77%) is the *p,p'* isomer (pictured above). The *o,p'* isomer (pictured to the right) is also present in significant amounts (15%). Dichlorodiphenyldichloroethylene (DDE) and dichlorodiphenyldichloroethane (DDD) make up the balance. DDE and DDD are the major metabolites and environmental breakdown products. The term "total DDT" is often used to refer to the sum of all DDT related compounds (*p,p'*-DDT, *o,p'*-DDT, DDE, and DDD) in a sample.

Production and Use

From 1950 to 1980, DDT was extensively used in agriculture — more than 40,000 tonnes each year worldwide — and it has been estimated that a total of 1.8 million tonnes have been produced globally since the 1940s. In the United States, it was manufactured by some 15 companies, including Monsanto, Ciba, Montrose Chemical Company, Pennwalt and Velsicol Chemical Corporation. Production peaked in 1963 at 82,000 tonnes per year. More than 600,000 tonnes (1.35 billion pounds) were applied in the US before the 1972 ban. Usage peaked in 1959 at about 36,000 tonnes.

In 2009, 3,314 tonnes were produced for malaria control and visceral leishmaniasis. India is the only country still manufacturing DDT and is the largest consumer. China ceased production in 2007.

Mechanism of Insecticide Action

In insects it opens sodium ion channels in neurons, causing them to fire spontaneously, which leads to spasms and eventual death. Insects with certain mutations in their sodium channel gene are resistant to DDT and similar insecticides. DDT resistance is also conferred by up-regulation of genes expressing cytochrome P450 in some insect species, as greater quantities of some enzymes of this group accelerate the toxin's metabolism into inactive metabolites.

History

DDT was first synthesized in 1874 by Othmar Zeidler under the supervision of Adolf von Baeyer. It was further described in 1929 in a dissertation by W. Bausch and in two subsequent publications in 1930. The insecticide properties of "multiple chlorinated aliphatic or fat-aromatic alcohols with at least one trichloromethane group" were described in a patent in 1934 by Wolfgang von Leuthold.

DDT's insecticidal properties were not, however, discovered until 1939 by the Swiss scientist Paul Hermann Müller, who was awarded the 1948 Nobel Prize in Physiology and Medicine for his efforts.

Commercial product concentrate containing 50% DDT, circa 1960s

Commercial product (Powder box, 50 g) containing 10% DDT; Néocide. Ciba Geigy DDT; *"Destroys parasites such as fleas, lice, ants, bedbugs, cockroaches, flies, etc.. Néocide Sprinkle caches of vermin and the places where there are insects and their places of passage. Leave the powder in place as long as possible.» «Destroy the parasites of man and his dwelling». «Death is not instantaneous, it follows inevitably sooner or later." "French manufacturing"; "harmless to humans and warm-blooded animals" "sure and lasting effect. Odorless."*

Use in The 1940S and 1950S

DDT is the best-known of several chlorine-containing pesticides used in the 1940s and 1950s. With pyrethrum in short supply, DDT was used extensively during World War II by the Allies to control the insect vectors of typhus – nearly eliminating the disease in many parts of Europe. In the South Pacific, it was sprayed aerially for malaria and dengue fever control with spectacular effects. While DDT's chemical and insecticidal properties were important factors in these victories, advances in application equipment coupled with competent organization and sufficient manpower were also crucial to the success of these programs.

In 1945, DDT was made available to farmers as an agricultural insecticide and played a role in the final elimination of malaria in Europe and North America.

In 1955, the World Health Organization commenced a program to eradicate malaria in countries with low to moderate transmission rates worldwide, relying largely on DDT for mosquito control and rapid diagnosis and treatment to reduce transmission. The program eliminated the disease in "Taiwan, much of the Caribbean, the Balkans, parts of northern Africa, the northern region of

Australia, and a large swath of the South Pacific" and dramatically reduced mortality in Sri Lanka and India.

However, failure to sustain the program, increasing mosquito tolerance to DDT, and increasing parasite tolerance led to a resurgence. In many areas early successes partially or completely reversed, and in some cases rates of transmission increased. The program succeeded in eliminating malaria only in areas with "high socio-economic status, well-organized healthcare systems, and relatively less intensive or seasonal malaria transmission".

DDT was less effective in tropical regions due to the continuous life cycle of mosquitoes and poor infrastructure. It was not applied at all in sub-Saharan Africa due to these perceived difficulties. Mortality rates in that area never declined to the same dramatic extent, and now constitute the bulk of malarial deaths worldwide, especially following the disease's resurgence as a result of resistance to drug treatments and the spread of the deadly malarial variant caused by *Plasmodium falciparum*.

Eradication was abandoned in 1969 and attention instead focused on controlling and treating the disease. Spraying programs (especially using DDT) were curtailed due to concerns over safety and environmental effects, as well as problems in administrative, managerial and financial implementation. Efforts shifted from spraying to the use of bednets impregnated with insecticides and other interventions.

United States Ban

As early as the 1940s, US scientists began expressing concern over possible hazards associated with DDT, and in the 1950s the government began tightening regulations governing its use. These events received little attention. In 1957 the *New York Times* reported an unsuccessful struggle to restrict DDT use in Nassau County, New York, that the issue came to the attention of the popular naturalist-author, Rachel Carson. William Shawn, editor of *The New Yorker*, urged her to write a piece on the subject, which developed into her 1962 book *Silent Spring*. The book argued that pesticides, including DDT, were poisoning both wildlife and the environment and were endangering human health. *Silent Spring* was a best seller, and public reaction to it launched the modern environmental movement in the United States. The year after it appeared, President John F. Kennedy ordered his Science Advisory Committee to investigate Carson's claims. The committee's report "add[ed] up to a fairly thorough-going vindication of Rachel Carson's Silent Spring thesis," in the words of the journal *Science*, and recommended a phaseout of "persistent toxic pesticides". DDT became a prime target of the growing anti-chemical and anti-pesticide movements, and in 1967 a group of scientists and lawyers founded the Environmental Defense Fund (EDF) with the specific goal of enacting a ban on DDT. Victor Yannacone, Charles Wurster, Art Cooley and others in the group had all witnessed bird kills or declines in bird populations and suspected that DDT was the cause. In their campaign against the chemical, EDF petitioned the government for a ban and filed lawsuits. Around this time, toxicologist David Peakall was measuring DDE levels in the eggs of peregrine falcons and California condors and finding that increased levels corresponded with thinner shells.

In response to an EDF suit, the U.S. District Court of Appeals in 1971 ordered the EPA to begin the de-registration procedure for DDT. After an initial six-month review process, William Ruckelshaus, the Agency's first Administrator rejected an immediate suspension of DDT's registration, citing studies from the EPA's internal staff stating that DDT was not an imminent danger. However, these findings were criticized, as they were performed mostly by economic entomologists inherited from the United States Department of Agriculture, who many environmentalists felt were biased towards agribusiness and understated concerns about human health and wildlife. The decision thus created controversy.

The EPA held seven months of hearings in 1971–1972, with scientists giving evidence for and against DDT. In the summer of 1972, Ruckelshaus announced the cancellation of most uses of DDT – exempting public health uses under some conditions. Immediately after the announcement, both EDF and the DDT manufacturers filed suit against EPA. Industry sought to overturn the ban, while EDF wanted a comprehensive ban. The cases were consolidated, and in 1973 the United States Court of Appeals for the District of Columbia Circuit ruled that the EPA had acted properly in banning DDT.

Some uses of DDT continued under the public health exemption. For example, in June 1979, the California Department of Health Services was permitted to use DDT to suppress flea vectors of bubonic plague. DDT continued to be produced in the United States for foreign markets until 1985, when over 300 tons were exported.

Restrictions on Usage

In the 1970s and 1980s, agricultural use was banned in most developed countries, beginning with Hungary in 1968 followed by Norway and Sweden in 1970, West Germany and the US in 1972, but not in the United Kingdom until 1984. By 1991 total bans, including for disease control, were in place in at least 26 countries; for example Cuba in 1970, Singapore in 1984, Chile in 1985 and the Republic of Korea in 1986.

The Stockholm Convention on Persistent Organic Pollutants, which took effect in 2004, outlawed several persistent organic pollutants, and restricted DDT use to vector control. The Convention was ratified by more than 170 countries. Recognizing that total elimination in many malaria-prone countries is currently unfeasible absent affordable/effective alternatives, the convention exempts public health use within World Health Organization (WHO) guidelines from the ban. Resolution 60.18 of the World Health Assembly commits WHO to the Stockholm Convention's aim of reducing and ultimately eliminating DDT. Malaria Foundation International states, "The outcome of the treaty is arguably better than the status quo going into the negotiations. For the first time, there is now an insecticide which is restricted to vector control only, meaning that the selection of resistant mosquitoes will be slower than before."

Despite the worldwide ban, agricultural use continued in India, North Korea, and possibly elsewhere as of 2008.

Today, about 3,000 to 4,000 tons of DDT are produced each year for disease vector control. DDT is applied to the inside walls of homes to kill or repel mosquitoes. This intervention, called indoor residual spraying (IRS), greatly reduces environmental damage. It also reduces the incidence of DDT resistance. For comparison, treating 40 hectares (99 acres) of cotton during a typical U.S. growing season requires the same amount of chemical as roughly 1,700 homes.

Environmental Impact

DDT is a persistent organic pollutant that is readily adsorbed to soils and sediments, which can act both as sinks and as long-term sources of exposure affecting organisms. Depending on conditions, its soil half life can range from 22 days to 30 years. Routes of loss and degradation include runoff, volatilization, photolysis and aerobic and anaerobic biodegradation. Due to hydrophobic properties, in aquatic ecosystems DDT and its metabolites are absorbed by aquatic organisms and adsorbed on suspended particles, leaving little DDT dissolved in the water. Its breakdown products and metabolites, DDE and DDD, are also persistent and have similar chemical and physical properties. DDT and its breakdown products are transported from warmer areas to the Arctic by the phenomenon of global distillation, where they then accumulate in the region's food web.

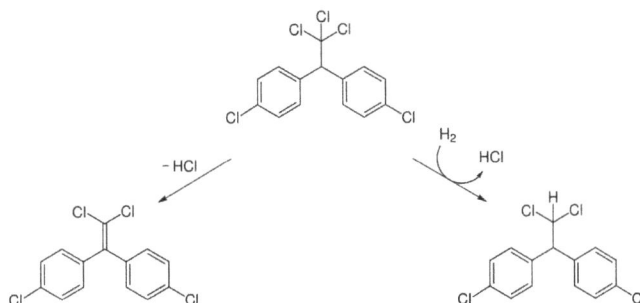

Degradation of DDT to form DDE (by elimination of HCl, left) and DDD (by reductive dechlorination, right)

Because of its lipophilic properties, DDT can bioaccumulate, especially in predatory birds. DDT, DDE and DDD magnify through the food chain, with apex predators such as raptor birds concentrating more chemicals than other animals in the same environment. They are stored mainly in body fat. DDT and DDE are resistant to metabolism; in humans, their half-lives are 6 and up to 10 years, respectively. In the United States, these chemicals were detected in almost all human blood samples tested by the Centers for Disease Control in 2005, though their levels have sharply declined since most uses were banned. Estimated dietary intake has declined, although FDA food tests commonly detect it.

Marine macroalgae (seaweed) help reduce soil toxicity by up to 80% within six weeks.

Effects on Wildlife and Eggshell Thinning

DDT is toxic to a wide range of living organisms, including marine animals such as crayfish, daphnids, sea shrimp and many species of fish. DDE caused eggshell thinning and population declines in multiple North American and European bird of prey species. Eggshell thinning lowers the reproductive success rate of certain bird species by causing egg breakage and embryo deaths. DDE-related eggshell thinning is considered a major reason for the decline of the bald eagle, brown pelican, peregrine falcon and osprey. However, birds vary in their sensitivity to these chemicals. Birds of prey, waterfowl and song birds are more susceptible than chickens and related species. DDE appears to be more potent than DDT. Even in 2010, California condors that feed on sea lions at Big Sur that in turn feed in the Palos Verdes Shelf area of the Montrose Chemical Superfund site exhibited continued thin-shell problems. Scientists with the Ventana Wildlife Society and others study and remediate the condors' problems.

The biological thinning mechanism is not entirely understood, but strong evidence indictates that p,p'-DDE inhibits calcium ATPase in the membrane of the shell gland and reduces the transport of calcium carbonate from blood into the eggshell gland. This results in a dose-dependent thickness reduction. Other evidence indicates that o,p'-DDT disrupts female reproductive tract development, later impairing eggshell quality. Multiple mechanisms may be at work, or different mechanisms may operate in different species. Some studies show that although DDE levels have fallen dramatically, eggshell thickness remains 10–12 percent thinner than before DDT was first used.

Human Health

DDT is an endocrine disruptor. It is considered likely to be a human carcinogen although the majority of studies suggest it is not directly genotoxic. DDE acts as a weak androgen receptor antagonist, but not as an estrogen. p,p'-DDT, DDT's main component, has little or no androgenic or estrogenic activity. The minor component o,p'-DDT has weak estrogenic activity.

A U.S. soldier is demonstrating DDT hand-spraying equipment. DDT was used
to control the spread of typhus-carrying lice.

Acute Toxicity

DDT is classified as "moderately toxic" by the US National Toxicology Program (NTP) and "moderately hazardous" by WHO, based on the rat oral LD_{50} of 113 mg/kg. DDT has on rare occasions been administered orally as a treatment for barbiturate poisoning.

Chronic Toxicity

DDT and DDE, like other organochlorines, have been shown to have xenoestrogenic activity, meaning they are chemically similar enough to estrogens to trigger hormonal responses in animals. This endocrine disrupting activity has been observed in mice and rat toxicological studies. Epidemiological evidence indicates that these effects may be occurring in humans as a result of DDT exposure. EPA states that DDT exposure damages the reproductive system and reduces reproductive success. These effects may cause developmental and reproductive toxicity:

- A review article in *The Lancet* states, "research has shown that exposure to DDT at amounts that would be needed in malaria control might cause preterm birth and early weaning ... toxicological evidence shows endocrine-disrupting properties; human data also indicate possible disruption in semen quality, menstruation, gestational length, and duration of lactation."

- Other studies document decreases in semen quality among men with high exposures (generally from IRS).

- Studies generally find that high blood DDT or DDE levels do not increase time to pregnancy (TTP.) Some evidence indicates that the daughters of highly exposed women may have more increased TTP.

- DDT is associated with early pregnancy loss, a type of miscarriage. A prospective cohort study of Chinese textile workers found "a positive, monotonic, exposure-response association between preconception serum total DDT and the risk of subsequent early pregnancy losses." The median serum DDE level of study group was lower than that typically observed in women living in homes sprayed with DDT.

- A Japanese study of congenital hypothyroidism concluded that *in utero* DDT exposure may affect thyroid hormone levels and "play an important role in the incidence and/or causation of cretinism." Other studies found that DDT or DDE interfere with proper thyroid function in pregnancy and childhood.

- Exposure to DDT can cause shorter menstrual cycles.

Carcinogenicity

In 2002, the Centers for Disease Control and Prevention reported, "Overall, in spite of some positive associations for some cancers within certain subgroups of people, there is no clear evidence that exposure to DDT/DDE causes cancer in humans." The NTP classifies it as "reasonably anticipated to be a carcinogen," the International Agency for Research on Cancer classifies it as "probably carcinogenic to humans", and the EPA classifies DDT, DDE and DDD as class B2 "probable" carcinogens. These evaluations are based mainly on animal studies.

A 2005 Lancet review stated that occupational DDT exposure was associated with increased pancreatic cancer risk in 2 case control studies, but another study showed no DDE dose-effect association. Results regarding a possible association with liver cancer and biliary tract cancer are conflicting: workers who did not have direct occupational DDT contact showed increased risk. White men had an increased risk, but not white women or black men. Results about an association with multiple myeloma, prostate and testicular cancer, endometrial cancer and colorectal cancer have been inconclusive or generally do not support an association.

A 2009 review, whose co-authors included persons engaged in DDT-related litigation, reached broadly similar conclusions, with an equivocal association with testicular cancer. Case–control studies did not support an association with leukemia or lymphoma.

Breast Cancer

The question of whether DDT or DDE are risk factors in breast cancer has not been conclusively answered. Several meta analyses of observational studies have concluded that there is no overall relationship between DDT exposure and breast cancer risk. The United States Institute of Medicine reviewed data on the association of breast cancer with DDT exposure in 2012 and concluded that a causative relationship could neither be proven nor disproven.

A 2007 case control study using archived blood samples found that breast cancer risk was increased 5-fold among women who were born prior to 1931 and who had high serum DDT levels in 1963. Reasoning that DDT use became widespread in 1945 and peaked around 1950, they concluded that the ages of 14-20 were a critical period in which DDT exposure leads to increased risk. This study, which suggests a connection between DDT exposure and breast cancer that would not be picked up by most studies, has received variable commentary in third party reviews. One review suggested that "previous studies that measured exposure in older women may have missed the critical period." A second review suggested a cautious approach to the interpretation of these results given methodological weaknesses in the study design. The National Toxicology Program notes that while the majority of studies have not found a relationship between DDT exposure and breast cancer that positive associations have been seen in a "few studies among women with higher levels of exposure and among certain subgroups of women"

A 2015 case control study identified a link (odds ratio 3.4) between *in-utero* exposure (as estimated from archived maternal blood samples) and breast cancer diagnosis in daughters. The findings "support classification of DDT as an endocrine disruptor, a predictor of breast cancer, and a marker of high risk".

Malaria

Malaria remains the primary public health challenge in many countries. 2008 WHO estimates were 243 million cases and 863,000 deaths. About 89% of these deaths occur in Africa, mostly to children under age 5. DDT is one of many tools to fight the disease. Its use in this context has been called everything from a "miracle weapon [that is] like Kryptonite to the mosquitoes," to "toxic colonialism".

Before DDT, eliminating mosquito breeding grounds by drainage or poisoning with Paris green or pyrethrum was sometimes successful. In parts of the world with rising living standards, the elimination of malaria was often a collateral benefit of the introduction of window screens and improved sanitation. A variety of usually simultaneous interventions represents best practice. These include antimalarial drugs to prevent or treat infection; improvements in public health infrastructure to diagnose, sequester and treat infected individuals; bednets and other methods intended to keep mosquitoes from biting humans; and vector control strategies such as larvaciding with insecticides, ecological controls such as draining mosquito breeding grounds or introducing fish to eat larvae and indoor residual spraying (IRS) with insecticides, possibly including DDT. IRS involves the treatment of interior walls and ceilings with insecticides. It is particularly effective against mosquitoes, since many species rest on an indoor wall before or after feeding. DDT is one of 12 WHO–approved IRS insecticides.

WHO's anti-malaria campaign of the 1950s and 1960s relied heavily on DDT and the results were promising, though temporary in developing countries. Experts tie malarial resurgence to multiple factors, including poor leadership, management and funding of malaria control programs; poverty; civil unrest; and increased irrigation. The evolution of resistance to first-generation drugs (e.g. chloroquine) and to insecticides exacerbated the situation. Resistance was largely fueled by unrestricted agricultural use. Resistance and the harm both to humans and the environment led many governments to curtail DDT use in vector control and agriculture. In 2006 WHO reversed a longstanding policy against DDT by recommending that it be used as an indoor pesticide in regions where malaria is a major problem.

Once the mainstay of anti-malaria campaigns, as of 2008 only 12 countries used DDT, including India and some southern African states, though the number was expected to rise.

Initial Effectiveness

When it was introduced in World War II, DDT was effective in reducing malaria morbidity and mortality. WHO's anti-malaria campaign, which consisted mostly of spraying DDT and rapid treatment and diagnosis to break the transmission cycle, was initially successful as well. For example, in Sri Lanka, the program reduced cases from about one million per year before spraying to just 18 in 1963 and 29 in 1964. Thereafter the program was halted to save money and malaria rebounded to 600,000 cases in 1968 and the first quarter of 1969. The country resumed DDT vector control but the mosquitoes had evolved resistance in the interim, presumably because of continued agricultural use. The program switched to malathion, but despite initial successes, malaria continued its resurgence into the 1980s.

DDT remains on WHO's list of insecticides recommended for IRS. After the appointment of Arata Kochi as head of its anti-malaria division, WHO's policy shifted from recommending IRS only in areas of seasonal or episodic transmission of malaria, to advocating it in areas of continuous, intense transmission. WHO reaffirmed its commitment to phasing out DDT, aiming "to achieve a 30% cut in the application of DDT world-wide by 2014 and its total phase-out by the early 2020s if not sooner" while simultaneously combating malaria. WHO plans to implement alternatives to DDT to achieve this goal.

South Africa continues to use DDT under WHO guidelines. In 1996, the country switched to alternative insecticides and malaria incidence increased dramatically. Returning to DDT and introducing new drugs brought malaria back under control. Malaria cases increased in South America after countries in that continent stopped using DDT. Research data showed a strong negative relationship between DDT residual house sprayings and malaria. In a research from 1993 to 1995, Ecuador increased its use of DDT and achieved a 61% reduction in malaria rates, while each of the other countries that gradually decreased its DDT use had large increases.

Mosquito Resistance

In some areas resistance reduced DDT's effectiveness. WHO guidelines require that absence of resistance must be confirmed before using the chemical. Resistance is largely due to agricultural use, in much greater quantities than required for disease prevention.

Resistance was noted early in spray campaigns. Paul Russell, former head of the Allied Anti-Malaria campaign, observed in 1956 that "resistance has appeared after six or seven years." Resistance has been detected in Sri Lanka, Pakistan, Turkey and Central America and it has largely been replaced by organophosphate or carbamate insecticides, *e.g.* malathion or bendiocarb.

In many parts of India, DDT is ineffective. Agricultural uses were banned in 1989 and its anti-malarial use has been declining. Urban use ended. DDT is still manufactured and used. One study concluded that "DDT is still a viable insecticide in indoor residual spraying owing to its effectivity in well supervised spray operation and high excito-repellency factor."

Studies of malaria-vector mosquitoes in KwaZulu-Natal Province, South Africa found susceptibility to 4% DDT (WHO's susceptibility standard), in 63% of the samples, compared to the average of 86.5% in the same species caught in the open. The authors concluded that "Finding DDT resistance in the vector *An. arabiensis*, close to the area where we previously reported pyrethroid-resistance in the vector *An. funestus* Giles, indicates an urgent need to develop a strategy of insecticide resistance management for the malaria control programmes of southern Africa."

DDT can still be effective against resistant mosquitoes and the avoidance of DDT-sprayed walls by mosquitoes is an additional benefit of the chemical. For example, a 2007 study reported that resistant mosquitoes avoided treated huts. The researchers argued that DDT was the best pesticide for use in IRS (even though it did not afford the most protection from mosquitoes out of the three test chemicals) because the others pesticides worked primarily by killing or irritating mosquitoes – encouraging the development of resistance. Others argue that the avoidance behavior slows eradication. Unlike other insecticides such as pyrethroids, DDT requires long exposure to accumulate a lethal dose; however its irritant property shortens contact periods. "For these reasons, when comparisons have been made, better malaria control has generally been achieved with pyrethroids than with DDT." In India outdoor sleeping and night duties are common, implying that "the excito-repellent effect of DDT, often reported useful in other countries, actually promotes outdoor transmission." Genomic studies in the model genetic organism *Drosophila melanogaster* revealed that high level DDT resistance is polygenic, involving multiple resistance mechanisms.

Residents' Concerns

IRS is effective if at least 80% of homes and barns in a residential area are sprayed. Lower coverage rates can jeopardize program effectiveness. Many residents resist DDT spraying, objecting to the lingering smell, stains on walls, and the potential exacerbation of problems with other insect pests. Pyrethroid insecticides (e.g. deltamethrin and lambda-cyhalothrin) can overcome some of these issues, increasing participation.

Human Exposure

A 1994 study found that South Africans living in sprayed homes have levels that are several orders of magnitude greater than others. Breast milk from South African mothers contains high levels of DDT and DDE. It is unclear to what extent these levels arise from home spraying vs food residues. Evidence indicates that these levels are associated with infant neurological abnormalities.

Most studies of DDT's human health effects have been conducted in developed countries where DDT is not used and exposure is relatively low.

Illegal diversion to agriculture is also a concern as it is difficult to prevent and its subsequent use on crops is uncontrolled. For example, DDT use is widespread in Indian agriculture, particularly mango production and is reportedly used by librarians to protect books. Other examples include Ethiopia, where DDT intended for malaria control is reportedly used in coffee production, and Ghana where it is used for fishing." The residues in crops at levels unacceptable for export have been an important factor in bans in several tropical countries. Adding to this problem is a lack of skilled personnel and management.

Criticism of Restrictions on DDT Use

Critics argue that limitations on DDT use for public health purposes have caused unnecessary morbidity and mortality from vector-borne diseases, with some claims of malaria deaths ranging as high as the hundreds of thousands and millions. Robert Gwadz of the US National Institutes of Health said in 2007, "The ban on DDT may have killed 20 million children." These arguments were rejected as "outrageous" by former WHO scientist Socrates Litsios. May Berenbaum, University of Illinois entomologist, says, "to blame environmentalists who oppose DDT for more deaths than Hitler is worse than irresponsible." Investigative journalist Adam Sarvana and others characterize this notion as a "myth" promoted principally by Roger Bate of the pro-DDT advocacy group Africa Fighting Malaria (AFM).

Criticisms of a DDT "ban" often specifically reference the 1972 United States ban (with the erroneous implication that this constituted a worldwide ban and prohibited use of DDT in vector control). Reference is often made to *Silent Spring,* even though Carson never pushed for a DDT ban. John Quiggin and Tim Lambert wrote, "the most striking feature of the claim against Carson is the ease with which it can be refuted."

It has been alleged that donor governments and agencies refused to fund DDT spraying, or made aid contingent upon not using DDT. According to a report in the *British Medical Journal*, use of DDT in Mozambique "was stopped several decades ago, because 80% of the country's health budget came from donor funds, and donors refused to allow the use of DDT." Roger Bate asserted, "many countries have been coming under pressure from international health and environment agencies to give up DDT or face losing aid grants: Belize and Bolivia are on record admitting they gave in to pressure on this issue from [USAID]."

The US Agency for International Development (USAID) has been the focus of much criticism. While the agency now funds DDT use in some African countries, in the past it did not. When John Stossel accused USAID of not funding DDT because it wasn't "politically correct," Anne Peterson, the agency's assistant administrator for global health, replied that "I believe that the strategies we are using are as effective as spraying with DDT ... So, politically correct or not, I am very confident that what we are doing is the right strategy." USAID's Kent R. Hill stated that the agency had been misrepresented: "USAID strongly supports spraying as a preventative measure for malaria and will support the use of DDT when it is scientifically sound and warranted." The Agency's website states that "USAID has never had a 'policy' as such either 'for' or 'against' DDT for IRS (Indoor residual spraying). The real change in the past two years [2006/07] was a new interest and emphasis on IRS in general – with DDT or any other insecticide – as an effective malaria prevention strategy in tropical Africa." The agency claimed that in many cases alternative malaria control measures were more cost-effective than DDT spraying.

Alternatives

Insecticides

Organophosphate and carbamate insecticides, *e.g.* malathion and bendiocarb, respectively, are more expensive than DDT per kilogram and are applied at roughly the same dosage. Pyrethroids such as deltamethrin are also more expensive than DDT, but are applied more sparingly (0.02–0.3 g/m^2 vs 1–2 g/m^2), so the net cost per house is about the same.

Non-chemical Vector Control

Before DDT, malaria was successfully eliminated or curtailed in several tropical areas by removing or poisoning mosquito breeding grounds and larva habitats, for example by eliminating standing water. These methods have seen little application in Africa for more than half a century. According to CDC, such methods are not practical in Africa because "*Anopheles gambiae*, one of the primary vectors of malaria in Africa, breeds in numerous small pools of water that form due to rainfall … It is difficult, if not impossible, to predict when and where the breeding sites will form, and to find and treat them before the adults emerge."

The relative effectiveness of IRS versus other malaria control techniques (e.g. bednets or prompt access to anti-malarial drugs) varies and is dependent on local conditions.

A WHO study released in January 2008 found that mass distribution of insecticide-treated mosquito nets and artemisinin–based drugs cut malaria deaths in half in malaria-burdened Rwanda and Ethiopia. IRS with DDT did not play an important role in mortality reduction in these countries.

Vietnam has enjoyed declining malaria cases and a 97% mortality reduction after switching in 1991 from a poorly funded DDT-based campaign to a program based on prompt treatment, bednets and pyrethroid group insecticides.

In Mexico, effective and affordable chemical and non-chemical strategies were so successful that the Mexican DDT manufacturing plant ceased production due to lack of demand.

A review of fourteen studies in sub-Saharan Africa, covering insecticide-treated nets, residual spraying, chemoprophylaxis for children, chemoprophylaxis or intermittent treatment for pregnant women, a hypothetical vaccine and changing front–line drug treatment, found decision making limited by the lack of information on the costs and effects of many interventions, the small number of cost-effectiveness analyses, the lack of evidence on the costs and effects of packages of measures and the problems in generalizing or comparing studies that relate to specific settings and use different methodologies and outcome measures. The two cost-effectiveness estimates of DDT residual spraying examined were not found to provide an accurate estimate of the cost-effectiveness of DDT spraying; the resulting estimates may not be good predictors of cost-effectiveness in current programs.

However, a study in Thailand found the cost per malaria case prevented of DDT spraying (US$1.87) to be 21% greater than the cost per case prevented of lambda-cyhalothrin–treated nets (US$1.54), casting some doubt on the assumption that DDT was the most cost-effective measure. The director of Mexico's malaria control program found similar results, declaring that it was 25% cheaper for Mexico to spray a house with synthetic pyrethroids than with DDT. However, another study in South Africa found generally lower costs for DDT spraying than for impregnated nets.

A more comprehensive approach to measuring cost-effectiveness or efficacy of malarial control would not only measure the cost in dollars, as well as the number of people saved, but would also consider ecological damage and negative human health impacts. One preliminary study found that it is likely that the detriment to human health approaches or exceeds the beneficial reductions in malarial cases, except perhaps in epidemics. It is similar to the earlier study regarding estimated theoretical infant mortality caused by DDT and subject to the criticism also mentioned earlier.

A study in the Solomon Islands found that "although impregnated bed nets cannot entirely replace DDT spraying without substantial increase in incidence, their use permits reduced DDT spraying."

A comparison of four successful programs against malaria in Brazil, India, Eritrea and Vietnam does not endorse any single strategy but instead states, "Common success factors included conducive country conditions, a targeted technical approach using a package of effective tools, data-driven decision-making, active leadership at all levels of government, involvement of communities, decentralized implementation and control of finances, skilled technical and managerial capacity at national and sub-national levels, hands-on technical and programmatic support from partner agencies, and sufficient and flexible financing."

DDT resistant mosquitoes have generally proved susceptible to pyrethroids. Thus far, pyrethroid resistance in *Anopheles* has not been a major problem.

Organophosphate

An organophosphate (sometimes abbreviated OP) or phosphate ester is the general name for esters of phosphoric acid. Many of the most important biochemicals are organophosphates, including DNA and RNA, as well as many of the cofactors essential for life. Organophosphates are the basis of many insecticides, herbicides, and nerve agents. The United States Environmental Protection Agency lists organophosphates as very highly acutely toxic to bees, wildlife, and humans. Recent studies suggest a possible link to adverse effects in the neurobehavioral development of fetuses and children, even at very low levels of exposure. Organophosphates are widely used as solvents, plasticizers, and EP additives.

Organophosphates are widely employed both in natural and synthetic applications because of the ease with which organic groups can be linked together.

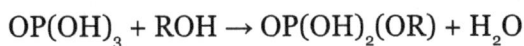

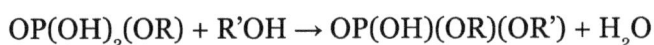

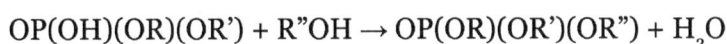

$$OP(OH)_3 + ROH \rightarrow OP(OH)_2(OR) + H_2O$$

$$OP(OH)_2(OR) + R'OH \rightarrow OP(OH)(OR)(OR') + H_2O$$

$$OP(OH)(OR)(OR') + R''OH \rightarrow OP(OR)(OR')(OR'') + H_2O$$

The phosphate esters bearing OH groups are acidic and partially deprotonated in aqueous solution. For example, DNA and RNA are polymers of the type $[PO_2(OR)(OR')^-]_n$. Polyphosphates also form esters; an important example of an ester of a polyphosphate is ATP, which is the monoester of triphosphoric acid ($H_5P_3O_{10}$).

Alcohols can be detached from phosphate esters by hydrolysis, which is the reverse of the above reactions. For this reason, phosphate esters are common carriers of organic groups in biosynthesis.

Organophosphate Pesticides

In health, agriculture, and government, the word "organophosphates" refers to a group of insecticides or nerve agents acting on the enzyme acetylcholinesterase (the pesticide group carbamates also act on this enzyme, but through a different mechanism). The term is used often to describe virtually any organic phosphorus(V)-containing compound, especially when dealing with neurotoxic compounds. Many of the so-called organophosphates contain C-P bonds. For instance, sarin is O-isopropyl methylphosphonofluoridate, which is formally derived from phosphorous acid ($HP(O)(OH)_2$), not phosphoric acid ($P(O)(OH)_3$). Also, many compounds which are derivatives of phosphinic acid are used as neurotoxic organophosphates.

Organophosphate pesticides (as well as sarin and VX nerve agent) irreversibly inactivate acetylcholinesterase, which is essential to nerve function in insects, humans, and many other animals. Organophosphate pesticides affect this enzyme in varied ways, and thus in their potential for poisoning. For instance, parathion, one of the first OPs commercialized, is many times more potent than malathion, an insecticide used in combatting the Mediterranean fruit fly (Med-fly) and West Nile Virus-transmitting mosquitoes.

Organophosphate pesticides degrade rapidly by hydrolysis on exposure to sunlight, air, and soil, although small amounts can be detected in food and drinking water. Their ability to degrade made them an attractive alternative to the persistent organochloride pesticides, such as DDT, aldrin, and dieldrin. Although organophosphates degrade faster than the organochlorides, they have greater acute toxicity, posing risks to people who may be exposed to large amounts.

Commonly used organophosphates have included parathion, malathion, methyl parathion, chlorpyrifos, diazinon, dichlorvos, phosmet, fenitrothion, tetrachlorvinphos, azamethiphos, and azinphos-methyl. Malathion is widely used in agriculture, residential landscaping, public recreation areas, and in public health pest control programs such as mosquito eradication. In the US, it is the most commonly used organophosphate insecticide. Forty organophosphate pesticides are registered in the U.S., with at least 73 million pounds used in agricultural and residential settings.

They are of concern to both scientists and regulators because they work by irreversibly blocking an enzyme critical to nerve function in both insects and humans. Even at relatively low levels, organophosphates may be most hazardous to the brain development of fetuses and young children. The EPA banned most residential uses of organophosphates in 2001, but they are still sprayed agriculturally on fruits and vegetables. They are also used to control pests such as mosquitos in public spaces such as parks. They can be absorbed through the lungs or skin or by eating them on food.

Organophosphates as Nerve Agents

History of Nerve Agents

Early pioneers in the field include Jean Louis Lassaigne (early 19th century) and Philippe de Clermont (1854). In 1932, German chemist Willy Lange and his graduate student, Gerde von Krueger, first described the cholinergic nervous system effects of organophosphates, noting a choking sensation and a dimming of vision after exposure. This discovery later inspired German chemist Gerhard Schrader at company IG Farben in the 1930s to experiment with these compounds as insecticides. Their potential use as chemical warfare agents soon became apparent, and the Nazi government put Schrader in charge of developing organophosphate (in the broader sense of the word) nerve gases. Schrader's laboratory discovered the G series of weapons, which included Sarin, Tabun, and Soman. The Nazis produced large quantities of these compounds, though did not use them during World War II. British scientists experimented with a cholinergic organophosphate of their own, called diisopropylfluorophosphate, during the war. The British later produced VX nerve agent, which was many times more potent than the G series, in the early 1950s, almost 20 years after the Germans had discovered the G series.

After World War II, American companies gained access to some information from Schrader's laboratory, and began synthesizing organophosphate pesticides in large quantities. Parathion was among the first marketed, followed by malathion and azinphosmethyl. The popularity of these insecticides increased after many of the organochlorine insecticides such as DDT, dieldrin, and heptachlor were banned in the 1970s.

Structural Features of Organophosphates

Effective organophosphates have the following structural features:

- A terminal oxygen connected to phosphorus by a double bond, i.e. a phosphoryl group

- Two lipophilic groups bonded to the phosphorus

- A leaving group bonded to the phosphorus, often a halide

Terminal Oxygen Vs. Terminal Sulfur

Thiophosphoryl compounds, those bearing the P=S functionality, are much less toxic than related phosphoryl derivatives. Thiophosphoryl compounds are not active inhibitors of acetylcholinesterase in either mammals or insects; in mammals, metabolism tends to remove lipophilic side groups from the phosphorus atom, while in insects it tends to oxidize the compound, thus removing the terminal sulfur and replacing it with a terminal oxygen, which allows the compound to more efficiently act as an acetylcholinesterase inhibitor.

Fine Tuning

Within these requirements, a large number of different lipophilic and leaving groups have been used. The variation of these groups is one means of fine tuning the toxicity of the compound. A good example of this chemistry are the *P*-thiocyanate compounds which use an aryl (or alkyl) group and an alkylamino group as the lipophilic groups. The thiocyanate is the leaving group.

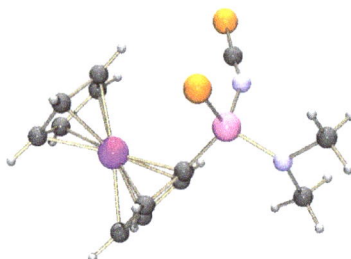

One of the products of the reaction of $Fc_2P_2S_4$ with dimethyl cyanamide

A German patent claimed that the reaction of 1,3,2,4-dithiadiphosphetane 2,4-disulfides with dialkyl cyanamides formed plant protection agents which contained six-membered (P-N=C-N=C-S-) rings. It has been proven in recent times by the reaction of diferrocenyl 1,3,2,4-dithiadiphosphetane 2,4-disulfide (and Lawesson's reagent) with dimethyl cyanamide that, in fact, a mixture of several different phosphorus-containing compounds is formed. Depending on the concentration of the dimethyl cyanamide in the reaction mixture, either a different six-membered ring compound (P-N=C-S-C=N-) or a nonheterocylic compound (FcP(S)(NR2)(NCS)) is formed as the major product; the other compound is formed as a minor product.

In addition, small traces of other compounds are also formed in the reaction. The ring compound (P-N=C-S-C=N-) {or its isomer} is unlikely to act as a plant protection agent, but (FcP(S)(NR2)(NCS)) compounds can act as nerve poisons in insects.

Organophosphate Poisoning

Many "organophosphates" are potent nerve agents, functioning by inhibiting the action of acetylcholinesterase (AChE) in nerve cells. They are one of the most common causes of poisoning worldwide, and are frequently intentionally used in suicides in agricultural areas. Organophosphate pesticides can be absorbed by all routes, including inhalation, ingestion, and dermal absorption. Their inhibitory effects on the acetylcholinesterase enzyme lead to a pathological excess of acetylcholine in the body. Their toxicity is not limited to the acute phase, however, and chronic effects have long been noted. Neurotransmitters such as acetylcholine (which is affected by organophosphate pesticides) are profoundly important in the brain's development, and many organophosphates have neurotoxic effects on developing organisms, even from low levels of exposure. Other organophosphates are not toxic, yet their main metabolites, such as their oxons, are. Treatment includes both a pralidoxime binder and an anticholinergic such as atropine.

Health Effects

Chronic Toxicity

Repeated or prolonged exposure to organophosphates may result in the same effects as acute exposure including the delayed symptoms. Other effects reported in workers repeatedly exposed include impaired memory and concentration, disorientation, severe depressions, irritability, confusion, headache, speech difficulties, delayed reaction times, nightmares, sleepwalking, drowsiness, or insomnia. An influenza-like condition with headache, nausea, weakness, loss of appetite, and malaise has also been reported.

Low-level Exposure

Even at relatively low levels, organophosphates may be hazardous to human health. The pesticides act on Acetylcholinesterase, an enzyme found in the brain chemicals closely related to those involved in ADHD, thus fetuses and young children, where brain development depends on a strict sequence of biological events, may be most at risk. They can be absorbed through the lungs or skin or by eating them on food. According to a 2008 report from the U.S. Department of Agriculture, "detectable" traces of organophosphate were found in a representative sample of produce tested by the agency, 28% of frozen blueberries, 20% of celery, 27% of green beans, 17% of peaches, 8% of broccoli, and 25% of strawberries.

The United States Environmental Protection Agency lists parathion as a possible human carcinogen.

A 2007 study linked the organophosphate insecticide chlorpyrifos, which is used on some fruits and vegetables, with delays in learning rates, reduced physical coordination, and behavioral problems in children, especially ADHD.

An organic diet is an effective way to reduce exposure to the organophosphorus pesticides commonly used in agricultural production. Organophosphate metabolite levels rapidly drop, and for some metabolites, become undetectable in children's urine when an organic diet is consumed.

Occupational organophosphate exposure is associated with an increased risk of Alzheimer's disease. Another study found that each 10-fold increase in urinary concentration of organophosphate metabolites was associated with a 55% to 72% increase in the odds of children being diagnosed with ADHD. Researchers analyzed the levels of organophosphate residues in the urine of more than 1,100 children aged 8 to 15 years old, and found that those with the highest levels of dialkyl phosphates, which are the breakdown products of organophosphate pesticides, also had the highest incidence of ADHD. Overall, they found a 35% increase in the odds of developing ADHD with every 10-fold increase in urinary concentration of the pesticide residues. The effect was seen even at the low end of exposure; children who had any detectable, above-average level of pesticide metabolite in their urine were twice as likely as those with undetectable levels to record symptoms of ADHD.

Children who were exposed to organophosphate pesticides while still in their mother's womb were more likely to develop attention disorders years later. Children at ages 3.5 and 5 years were evaluated for symptoms of attention disorders and ADHD using maternal reports of child behavior, performance on standardized computer tests, and behavior ratings from examiners. Each 10-fold increase in prenatal pesticide metabolites was linked to having five times the odds of scoring high on the computerized tests at age 5, suggesting a greater likelihood of a child having ADHD. The effect appeared to be stronger for boys than for girls.

Prenatal organophosphate exposure had a significant impact on birthweight and gestational age. A 10-fold increase in organophosphates concentrations in the mother was associated with a 0.5-week decrease in the infant's gestational age and a birth weight decline of 151 g (adjusted to account for changes in gestational age). "Diet and home pesticide use have been identified as important routes of exposure in non-agricultural populations," the researchers wrote, but noted that switching children from conventional to organic diets for several days reduced levels near or below the limit of detection, "suggesting that diet was the primary source of exposure in that study population."

Proposal Restrictions

According to the nongovernmental organisation Pesticide Action Network, parathion is one of the most dangerous pesticides. In the US alone, more than 650 agricultural workers have been poisoned since 1966, of which 100 died. In underdeveloped countries, many more people have suffered fatal and nonfatal intoxications. The World Health Organization, PAN, and numerous environmental organisations propose a general and global ban. Its use is banned or restricted in 23 countries and its import is illegal in a total of 50 countries. Its use was banned in the U.S. in 2000 and it has not been used since 2003.

Other than for agricultural use, the organophosphate diazinon has been banned in the U.S. More than one million pounds of diazinon were used in California to control agricultural pests in 2000. The areas and crops on which diazinon are most heavily applied are structural pest control, almonds, head lettuce, leaf lettuce, and prunes.

In May 2006, the Environmental Protection Agency (EPA) reviewed the use of dichlorvos and proposed its continued sale, despite concerns over its safety and considerable evidence suggesting it is carcinogenic and harmful to the brain and nervous system, especially in children. Environmentalists charge that the latest decision was the product of backroom deals with industry and political interference.

In 2001, the EPA placed new restrictions on the use of the organophosphates phosmet and azinphos-methyl to increase protection of agricultural workers. The crop uses reported at that time as being phased out in four years included those for almonds, tart cherries, cotton, cranberries, peaches, pistachios, and walnuts. The crops with time-limited registration included apples/crab apples, blueberries, sweet cherries, pears, pine seed orchards, brussels sprouts, cane berries, and the use of azinphos-methyl by nurseries for quarantine requirements. The labeled uses of phosmet include alfalfa, orchard crops (e.g. almonds, walnuts, apples, cherries), blueberries, citrus, grapes, ornamental trees (not for use in residential, park, or recreational areas) and nonbearing fruit trees, Christmas trees and conifers (tree farms), potatoes, and peas. Azinphos-methyl has been banned in Europe since 2006.

Neonicotinoid

Neonicotinoids (sometimes shortened to neonics) are a class of neuro-active insecticides chemically similar to nicotine. In the 1980s Shell and in the 1990s Bayer started work on their development. The neonicotinoid family includes acetamiprid, clothianidin, imidacloprid, nitenpyram, nithiazine, thiacloprid and thiamethoxam. Imidacloprid is the most widely used insecticide in the world. Compared to organophosphate and carbamate insecticides neonicotinoids cause less toxicity in birds and mammals than insects. Some breakdown products are also toxic.

In the late 1990s neonicotinoids came under increasing scrutiny over their environmental impacts. Neonicotinoid use was linked in a range of studies to adverse ecological effects, including honey-bee colony collapse disorder (CCD) and loss of birds due to a reduction in insect populations. In 2013, the European Union and a few non EU countries restricted the use of certain neonicotinoids.

History

The precursor to nithiazine was first synthesized by Henry Feuer, a chemist at Purdue University, in 1970; Shell researchers found in screening that this precursor showed insecticide potential and refined it to develop nithiazine. In 1984 nithiazine's mode of action was found to be as a postsynaptic acetylcholine receptor agonist, the same as nicotine. Nithiazine does not act as an acetylcholinesterase inhibitor, in contrast to the organophosphate and carbamate insecticides. While nithiazine has the desired specificity (i.e. low mammalian toxicity), it is not photostable—that is, it breaks down in sunlight, and thus is not commercially viable.

In 1985, Bayer patented imidacloprid as the first commercial neonicotinoid.

During the late 1990s, primarily, imidacloprid became widely used. Beginning in the early 2000s, two other neonicotinoids, clothianidin and thiamethoxam entered the market. As of 2013, virtually all corn planted in the United States was treated with one of these two insecticides and various fungicides. As of 2014, about a third of US soybean acreage was planted with neonicotinoid treated seeds, usually imidacloprid or thiamethoxam.

Market

Neonicotinoids have been registered in more than 120 countries. With a global turnover of €1.5 billion in 2008, they represented 24% of the global market for insecticides. After the introduction of the first neonicotinoids in the 1990s, this market has grown from €155 million in 1990 to €957 million in 2008. Neonicotinoids made up 80% of all seed treatment sales in 2008.

As of 2011, seven neonicotinoids from different companies are on the market.

Name	Company	Products	Turnover in million US$ (2009)
Imidacloprid	Bayer CropScience	Confidor, Admire, Gaucho, Advocate	1,091
Thiamethoxam	Syngenta	Actara, Platinum, Cruiser	627
Clothianidin	Sumitomo Chemical/Bayer CropScience	Poncho, Dantosu, Dantop	439
Acetamiprid	Nippon Soda	Mospilan, Assail, ChipcoTristar	276
Thiacloprid	Bayer CropScience	Calypso	112
Dinotefuran	Mitsui Chemicals	Starkle, Safari, Venom	79
Nitenpyram	Sumitomo Chemical	Capstar, Guardian	8

Agricultural Usage

Efficacy

Imidacloprid is effective against sucking insects, some chewing insects, soil insects and fleas on domestic animals. It is systemic with particular efficacy against sucking insects and has a long residual activity. Imidacloprid can be added to the water used to irrigate plants. Controlled release formulations of imidacloprid take 2–10 days to release 50% of imidacloprid in water. It is applied against soil pests, seed, timber and animal pests as well as foliar treatments.

As of 2013 neonicotinoids have been used In the U.S. on about 95 percent of corn and canola crops, the majority of cotton, sorghum, and sugar beets and about half of all soybeans. They have been used on the vast majority of fruit and vegetables, including apples, cherries, peaches, oranges, berries, leafy greens, tomatoes, and potatoes, to cereal grains, rice, nuts, and wine grapes. Imidacloprid is possibly the most widely used insecticide, both within the neonicotinoids and in the worldwide market.

Seed Coatings

In agriculture, usefulness of neonicotinoid seed treatments for pest prevention depends upon the timing of planting and pest arrival. For soybeans, neonicotinoid seed treatments typically are not effective against the soybean aphid, because the compounds break down 35–42 days after planting, and soybean aphids typically are not present or at damaging population levels before this time. Neonicotinoid seed treatments can protect yield in special cases such as late-planted fields or in areas with large infestations much earlier in the growing season. Overall yield gains are not expected from neonicotinoid seed treatments for soybean insect pests in the United States, and foliar insecticides are recommended instead when insects do reach damaging levels. Health Canada estimated that neonicotinoids provide benefits equivalent to over 3% of the national farm gate value of corn and 1.5% to 2.1% of the national farm gate value of soybean in 2013 .

Regulation

United States

The US EPA operates a 15-year registration review cycle for all pesticides. The EPA granted a conditional registration to clothianidin in 2003. The EPA issues conditional registrations when a pesticide meets the standard for registration, but there are outstanding data requirements. Thiamethoxam is approved for use as an antimicrobial pesticide wood preservative and as a pesticide; it was first approved in 1999. Imidacloprid was registered in 1994.

As all neonicotinoids were registered after 1984, they were not subject to reregistration, but due to environmental concerns, especially concerning bees, the EPA opened dockets to evaluate them. The registration review docket for imidacloprid opened in December 2008, and the docket for nithiazine opened in March 2009. To best take advantage of new research as it becomes available, the EPA moved ahead the docket openings for the remaining neonicotinoids on the registration review schedule (acetamiprid, clothianidin, dinotefuran, thiacloprid, and thiamethoxam) to FY 2012. The EPA has said that it expects to complete the review for the neonicotinoids in 2018.

In March 2012, the Center for Food Safety, Pesticide Action Network, Beyond Pesticides and a group of beekeepers filed an Emergency Petition with the EPA asking the agency to suspend the use of clothianidin. The agency denied the petition. In March 2013, the US EPA was sued by the same group, with the Sierra Club and the Center for Environmental Health joining, which accused the agency of performing inadequate toxicity evaluations and allowing insecticide registration based on inadequate studies. The case, *Ellis et al v. Bradbury et al*, was stayed as of October 2013.

On July 12, 2013, Rep. John Conyers, on behalf of himself and Rep. Earl Blumenauer, introduced the "Save American Pollinators Act" in the House of Representatives. The Act called for suspension of the use of four neonicotinoids, including the three recently suspended by the European Union, until their review is complete, and for a joint Interior Department and EPA study of bee populations and the possible reasons for their decline. The bill was assigned to a congressional committee on July 16, 2013 and did not leave committee.

Europe

In 2008, Germany revoked the registration of clothianidin for use on seed corn after an incident that resulted in the death of millions of nearby honey bees. An investigation revealed that it was caused by a combination of factors:

- failure to use a polymer seed coating known as a "sticker"

- weather conditions that resulted in late planting when nearby canola crops were in bloom;

- a particular type of air-driven equipment used to sow the seeds which apparently blew clothianidin-laden dust off the seeds and into the air as the seeds were ejected from the machine into the ground;

- dry and windy conditions at the time of planting that blew the dust into the nearby canola fields where honey bees were foraging;

In Germany, clothianidin use was also restricted in 2008 for a short period on rapeseed. After it was shown that rapeseed treatment did not have the same problems as maize, its use was reinstated under the condition that the pesticide be fixed to the rapeseed grains by an additional sticker, so that abrasion dusts would not be released into the air.

In 2009, the German Federal Office of Consumer Protection and Food Safety decided to continue to suspend authorization for clothianidin use on corn. It had not yet been fully clarified to what extent and in what manner bees come into contact with the active substances in clothianidin, thiamethoxam and imidacloprid when used on corn. The question of whether liquid emitted by plants via guttation, which bees ingest, posed an additional risk was unanswered.

Neonicotinoid seed treatment is banned in Italy, but foliar use is allowed. This action was taken based on preliminary monitoring studies showing that bee losses were correlated with the application of seeds treated with these compounds; Italy based its decision on the known acute toxicity of these compounds to pollinators.

In France, sunflower and corn seed treatment with imidacloprid are suspended; imidacloprid seed treatment for sugar beets and cereals are allowed, as is foliar use.

In 2012, the European Commission asked the European Food Safety Authority (EFSA) to study the safety of three neonicotinoids, in response to growing concerns about the impact of neonicotinoids on honey bees. The study was published in January 2013, stating that neonicotinoids pose an unacceptably high risk to bees, and that the industry-sponsored science upon which regulatory agencies' claims of safety have relied may be flawed and contain data gaps not previously considered. Their review concluded, "A high acute risk to honey bees was identified from exposure via dust drift for the seed treatment uses in maize, oilseed rape and cereals. A high acute risk was also identified from exposure via residues in nectar and/or pollen." EFSA reached the following conclusions:

- Exposure from pollen and nectar. Only uses on crops not attractive to honey bees were considered acceptable.

- Exposure from dust. A risk to honey bees was indicated or could not be excluded, with some exceptions, such as use on sugar beet and crops planted in glasshouses, and for the use of some granules.

- Exposure from guttation. The only completed assessment was for maize treated with thiamethoxam. In this case, field studies showed an acute effect on honey bees exposed to the substance through guttation fluid.

EFSA's scientists identified a number of data gaps and were unable to finalize risk assessments for some uses authorized in the EU. EFSA also highlighted that risk to other pollinators should be further considered. The UK Parliament asked manufacturer Bayer Cropscience to explain discrepancies in the evidence they submitted.

In response to the study, the European Commission recommended a restriction of their use across the European Union. On 29 April 2013, 15 of the 27 EU member states voted to restrict the use of three neonicotinoids for two years starting 1 December 2013. Eight nations voted against the ban, while four abstained. The law restricts the use of imidacloprid, clothianidin and thiamethoxam for seed treatment, soil application (granules) and foliar treatment in crops attractive to bees. Temporary suspensions had previously been enacted in France, Germany and Italy. In Switzerland, where neonicotinoids were never used in alpine areas, neonics were banned due to accidental poisonings of bee populations and the relatively low safety margin for other beneficial insects.

Environmentalists called the move "a significant victory for common sense and our beleaguered bee populations" and said it is "crystal clear that there is overwhelming scientific, political and public support for a ban." The UK, which voted against the bill, disagreed: "Having a healthy bee population is a top priority for us, but we did not support the proposal for a ban because our scientific evidence doesn't support it." Bayer Cropscience, which makes two of the three banned products, remarked "Bayer remains convinced neonicotinoids are safe for bees, when used responsibly and properly ... clear scientific evidence has taken a back-seat in the decision-making process." Reaction in the scientific community was mixed. Biochemist Lin Field said the decision was based on "political lobbying" and could lead to the overlooking of other factors involved in colony collapse disorder. Zoologist Lynn Dicks of Cambridge University disagreed, saying "This is a victory for the precautionary principle, which is supposed to underlie environmental regulation." Simon Potts, Professor of Biodiversity and Ecosystem Services at Reading University, called the ban "excellent news for pollinators", and said, "The weight of evidence from researchers clearly points to the need to have a phased ban of neonicotinoids."

Economic Impact

In January 2013, the Humboldt Forum for Food and Agriculture e. V. (HFFA), a non-profit think tank, published a report on the value of neonicotinoids in the EU. At their website HFFA lists as their partners/supporters: BASF FE, the world's largest chemical company; Bayer CropScience, makers of products for crop protection and nonagricultural pest control; E.ON, an electric utility service provider; KWS Seed, a seed producer; and the food company Nestlé.

The study was supported by COPA-COGECA, the European Seed Association and the European Crop Protection Association, and financed by neonicotinoid manufacturers Bayer CropScience and Syngenta. The report looked at the short- and medium-term impacts of a complete ban of all neonicotinoids on agricultural and total value added (VA) and employment, global prices, land use and greenhouse gas (GHG) emissions. In the first year, agricultural and total VA would decline by €2.8 and €3.8 billion, respectively. The greatest losses would be in wheat, maize and rapeseed in the UK, Germany, Romania and France. 22,000 jobs would be lost, primarily in Romania and Poland, and agricultural incomes would decrease by 4.7%. In the medium-term (5-year ban), losses would amount to €17 billion in VA, and 27,000 jobs. The greatest income losses would affect the UK, while most jobs losses would occur in Romania. Following a ban, the lowered production would induce more imports of agricultural commodities into the EU. Agricultural production outside the EU would expand by 3.3 million hectares, leading to additional emissions of 600 million tons of carbon dioxide equivalent.

When the report was released, Peter Melchett, policy director of the Soil Association, which has been working to ban neonicotinoids in the UK, commented that since the report was funded by Bayer Crop Sciences and Syngenta, "it was probably unlikely to conclude that neonicotinoids should be banned". The spokesperson further stated: "On the one hand, the chemical companies say we risk the additional costs to farmers amounting to £630 million. On the other, the possible cost of losing pollinating insects is thought to be worth three times as much (£1.8 billion*) to UK farmers."

Canada

In Montreal, Quebec, nearly all corn seeds and a majority of soybeans get treated with neonicotinoids. In the summer of 2015, Montreal passed a law to reduce the presence of neonicotinoids. Montreal's regulations were written to reduce the percent of seeds and beans covered with neonicotoids to 20 percent within two years.

Agricultural businesses oppose Montreal's ban. CropLife Canada is a trade association that represents manufacturers of agricultural plant science and pest management products. The main argument against Montreal's ban is that once farmers are no longer allowed to use neonicotinoids, they would likely use dangerous pesticide sprays on seeds. The industry's opposition centers around a White House pollinator health task force report and a Canadian Senate report. The reports said that bees face more serious threats than "scientifically safe neonics."

On December 10, 2015, Montreal banned all neonicotinoids - without exception - on all properties within the city limits, including the Botanical Garden, all agricultural areas and all golf courses.

Mode of Action

Neonicotinoids, like nicotine, bind to nicotinic acetylcholine receptors (nAChRs) of a cell and trigger a response by that cell. In mammals, nicotinic acetylcholine receptors are located in cells of both the central nervous system and peripheral nervous systems. In insects these receptors are limited to the central nervous system. Nicotinic acetylcholine receptors are activated by the neurotransmitter acetylcholine. While low to moderate activation of these receptors causes nervous stimulation, high levels overstimulate and block the receptors, causing paralysis and death. Acetylcholinesterase breaks down acetylcholine to terminate signals from these receptors. However, acetylcholinesterase cannot break down neonicotinoids and their binding is irreversible.

Basis of Selectivity

Mammals and insects have different composition of the receptor subunits and the structures of the receptors. Because most neonicotinoids bind much more strongly to insect neuron receptors than to mammal neuron receptors, these insecticides are more toxic to insects than mammals.

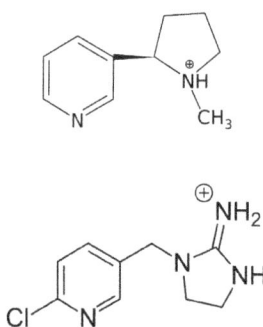

R-nicotine (top) and desnitro-imidacloprid are both protonated in the body

The low mammalian toxicity of imidacloprid has been explained by its inability to cross the blood–brain barrier because of lack of a charged nitrogen atom at physiological pH. The uncharged molecule can penetrate the insect blood–brain barrier.

Other neonicotinoids have a negatively charged nitro or cyano group, which interacts with a unique, positively charged amino acid residue present on insect, but not mammalian nAChRs.

However, the breakdown product desnitro-imidacloprid, which is formed in a mammal's body during metabolism as well as in environmental breakdown, has a charged nitrogen and shows high affinity to mammalian nAChRs. Desnitro-imidacloprid is quite toxic to mice.

Chemical Properties

Most neonicotinoids are water-soluble and break down slowly in the environment, so they can be taken up by the plant and provide protection from insects as the plant grows. Independent studies show that the photodegradation half-life time of most neonicotinoids is around 34 days when exposed to sunlight. However, it might take up to 1,386 days (3.8 years) for these compounds to degrade in the absence of sunlight and micro-organism activity. Some researchers are concerned that neonicotinoids applied agriculturally might accumulate in aquifers.

Toxicity

Decline in Bee Population

A dramatic rise in the number of annual beehive losses noticed around 2006 spurred interest in factors potentially affecting bee health. When first introduced, neonicotinoids were thought to have low toxicity to many insects, but recent research has suggested a potential toxicity to honey bees and other beneficial insects even with low levels of contact. Neonicotinoids may impact bees' ability to forage, learn and remember navigation routes to and from food sources. Separate from lethal and sublethal effects solely due to neonicotinoid exposure, neonicotinoids are also being explored with a combination with other factors, such as mites and pathogens, as potential causes of colony collapse disorder. Neonicotinoids may be responsible for detrimental effects on bumble bee colony growth and queen production. For example, *Bombus affinis*, a bumblebee endemic to North America, has decreased in nearly 90% of its natural habitats, much of which has been attributed to the use of neonicotoid based pesticides.

Previously undetected routes of exposure for bees include particulate matter or dust, pollen and nectar. Bees can fail to return to the hive without immediate lethality due to sub-nanogram toxicity, one primary symptom of colony collapse disorder. Separate research showed environmental persistence in agricultural irrigation channels and soil.

A 2012 study showed the presence of thiamethoxam and clothianidin in bees found dead in and around hives situated near agricultural fields. Other bees at the hives exhibited tremors, uncoordinated movement and convulsions, all signs of insecticide poisoning. The insecticides were also consistently found at low levels in soil up to two years after treated seed was planted and on nearby dandelion flowers and in corn pollen gathered by the bees. Insecticide-treated seeds are covered with a sticky substance to control its release into the environment, however they are then coated with talc to facilitate machine planting. This talc may be released into the environment in large amounts. Exhausted talc containing the insecticides is concentrated enough that even small amounts on flowering plants can kill foragers or be transported to the hive in contaminated pollen. Tests also showed that the corn pollen that bees were bringing back to hives tested positive for neonicotinoids at levels roughly below 100 parts per billion, an amount not acutely toxic, but enough to kill bees if sufficient amounts are consumed.

A 2013 review concluded that neonicotinoids as they are typically used harm bees and that safer alternatives are urgently needed. An October 2013 study by Italian researchers demonstrated that neonicotinoids disrupt bees' immune systems, making them susceptible to viral infections to which the bees are normally resistant.

In April 2015 EASAC conducted a study of the potential effects on organisms providing a range of ecosystem services like pollination and natural pest control which are critical to sustainable agriculture. The resulting report concludes "there is an increasing body of evidence that the widespread prophylactic use of neonicotinoids has severe negative effects on non-target organisms that provide ecosystem services including pollination and natural pest control." Two studies published in *Nature* provided further evidence of the deleterious effect of neonicontinoids on bees, although the further research is needed to corroborate the findings: Oilseed rape seed coated with a combination of clothianidin and a pyrethroid "reduced wild bee density, solitary bee nesting, and bumblebee colony growth and reproduction under field conditions". In a feeding experiment, bees preferred sucrose solutions with imidacloprid or thiamethoxam, even though it "caused them to eat less food overall".

An October 2015 study demonstrates significant effects on the survival and reproductive capacities of honey bee queens exposed to neonicotinoids. Those exposed to neonicotinoids had 60% survival rates, as compared to 80% for control groups. Lower worker egg production and alterations to surviving queens' reproductive anatomy "likely corresponded to reduced queen success (alive and producing worker offspring)." The authors further claim "our study suggests that these substances [i.e., neonicotinoids] are, at least partially, responsible for harming queens and causing population declines of social bee species. Failure of queens exposed to neonicotinoids during development to successfully lay fertilised eggs that subsequently develop into workers or queens is worrisome; both castes are vital to colony survival..."

Other Wildlife

In March 2013, the American Bird Conservancy published a commentary on 200 studies on neonicotinoids calling for a ban on neonicotinoid use as seed treatments because of their toxicity to birds, aquatic invertebrates, and other wildlife.

A 2013 Dutch study found that water containing allowable concentrations of imidacloprid had 50% fewer invertebrate species compared with uncontaminated water. A later study found the analysis was confounded with other co-occurring insecticides and did not show imidacloprid directly affected invertebrate diversity.

In the July 2014 issue of the journal *Nature*, a study based on an observed correlation between declines in some bird populations and the use of neonicotinoid pesticides in the Netherlands demonstrated that the level of neonicotinoids detected in environmental samples correlated strongly with the decline in populations of insect-eating birds. An editorial published in the same edition found the possible link between neonicotinoid pesticide use and a decline in bird numbers "worrying", pointing out that the persistence of the compounds (half-life of 1000 days) and the low direct toxicity to birds themselves implies that the depletion of the birds' food source (insects) is likely responsible for the decline and that the compounds are distributed widely in the environment. The editors write that while correlation is not the same as causation, "the authors of the study also rule out confounding effects from other land-use changes or pre-existing trends in bird declines".

From June to October 2014 a comprehensive Worldwide Integrated Assessment of the impact of Systemic Pesticides on biodiversity and ecosystems (WIA) was published in the journal Environmental Science and Pollution Research. In a series of papers it concludes that these systemic insecticides pose a serious risk of harm to a broad range of non-target invertebrate taxa often below the expected environmental concentrations. Their present scale use is therefore not a sustainable pest management approach and compromises the actions of numerous stakeholders in maintaining and supporting biodiversity and subsequently the ecological functions and services the diverse organisms perform.

Thiamethoxam

Thiamethoxam is a systemic insecticide in the class of neonicotinoids. It has a broad spectrum of activity against many types of insects.

History

Thiamethoxam was developed by Syngenta; a patent dispute arose with Bayer which already had patents covering other neonicotinoids including imidacloprid. In 2002 the dispute was settled, with Syngenta paying Bayer $120 million in exchange for worldwide rights to thiamethoxam.

Mechanisms of Action

Thiamethoxam is a broad-spectrum, systemic insecticide, which means it is absorbed quickly by plants and transported to all of its parts, including pollen, where it acts to deter insect feeding. An insect can absorb it in its stomach after feeding, or through direct contact, including through its tracheal system. The compound gets in the way of information transfer between nerve cells by interfering with nicotinic acetylcholine receptors in the central nervous system, and eventually paralyzes the muscles of the insects.

Syngenta asserts that thiamethoxam improves plant vigor by triggering physiological reactions within the plant, which induce the expression of specific "functional proteins" involved in various stress defense mechanisms of the plant allowing it to better cope under tough growing conditions, such as "drought and heat stress leading to protein degradation, low pH, high soil salinity, free radicals from UV radiation, toxic levels of aluminum, wounding from pests, wind, hail, etc, virus attack".

Toxicity

The selective toxicity of neonicotinoids like thiamethoxam for insects versus mammals is due to the higher sensitivity of insects' acetylcholine receptors.

The Food and Agriculture Organization (FAO) of the U.N. assessed thiamethoxam as "moderately hazardous to humans (WHO class III)", because it is harmful if swallowed. It found it to be no skin or eye irritant, and not mutagenic in any *in vitro* and *in vivo* toxicology tests.

FAO described thiamethoxam as non-toxic to fish, daphnia and algae, mildly toxic for birds, highly toxic to midges and acutely toxic for bees. The Globally Harmonized System of Classification and Labelling of Chemicals (GHS) classification is: "Harmful if swallowed. Very toxic to aquatic life with long lasting effects".

Sublethal doses of thiamethoxam metabolite clothianidin (0.05–2 ng/bee) have been known to cause reduced foraging activity since at least 1999, but this was quantified in 2012 by RFID tagged honeybees. Doses of equal or more than 0.5 ng/bee caused longer foraging flights.

Regulation

USA

Thiamethoxam is approved for use in the US as an antimicrobial pesticide wood preservative and as a pesticide; it was first approved in 1999. As of 2014, it is approved for use in a wide range of crops.

On September 5, 2014 Syngenta petitioned the EPA to increase the legal tolerance for thiamethoxam residue in numerous crops. It wants to use thiamethoxam as a leaf spray, rather than just a seed treatment, to treat late to midseason insect pests.

European Union Ban in 2013

In 2012, several peer reviewed independent studies were published showing that several neonicotinoids had previously undetected routes of exposure affecting bees including through dust, pollen, and nectar; that sub-nanogram toxicity resulted in failure to return to the hive without immediate lethality, the primary symptom of colony collapse disorder; and showing environmental persistence in agricultural irrigation channels and soil. However, not all studies have found significant effects, and the studies carried out to date have not reached a clear conclusion on the impacts of neonicotinoids. These reports prompted a formal peer review by the European Food Safety Authority, which stated in January 2013 that neonicotinoids pose an unacceptably high risk to bees, and that the industry-sponsored science upon which regulatory agencies' claims of safety have relied on may be flawed and contain several data gaps not previously considered. In April 2013, the European Union voted for a two-year restriction on neonicotinoid insecticides. The ban restricts the use of imidacloprid, clothianidin, and thiamethoxam on crops that attract bees.

Other Countries

Thiamethoxam is approved for a wide range of agricultural, viticultural(vineyard), and horticultural uses.

Imidacloprid

Imidacloprid is a systemic insecticide which acts as an insect neurotoxin and belongs to a class of chemicals called the neonicotinoids which act on the central nervous system of insects, with much lower toxicity to mammals. The chemical works by interfering with the transmission of stimuli in the insect nervous system. Specifically, it causes a blockage of the nicotinergic neuronal pathway. By blocking nicotinic acetylcholine receptors, imidacloprid prevents acetylcholine from transmitting impulses between nerves, resulting in the insect's paralysis and eventual death. It is effective on contact and via stomach action. Because imidacloprid binds much more strongly to insect neuron receptors than to mammal neuron receptors, this insecticide is more toxic to insects than to mammals.

As of 1999, Imidacloprid was the most widely used insecticide in the world. Although it is now off patent, the primary manufacturer of this chemical is Bayer CropScience (part of Bayer AG). It is sold under many names for many uses; it can be applied by soil injection, tree injection, application to the skin of the plant, broadcast foliar, ground application as a granular or liquid formulation, or as a pesticide-coated seed treatment. Imidacloprid is widely used for pest control in agriculture. Other uses include application to foundations to prevent termite damage, pest control for gardens and turf, treatment of domestic pets to control fleas, protection of trees from boring insects, and in preservative treatment of some types of lumber products (e.g., Ecolife brand).

Recent research suggests that widespread agricultural use of imidacloprid and other pesticides may be contributing to honey bee colony collapse disorder, the decline of honey bee colonies in Europe and North America observed since 2006. As a result, several countries have restricted use of imidacloprid and other neonicotinoids. In January 2013, the European Food Safety Authority stated that neonicotinoids pose an unacceptably high risk to bees, and that the industry-sponsored science upon which regulatory agencies' claims of safety have relied, may be flawed, or even deceptive.

Authorized Uses

Imidacloprid is the most widely used insecticide in the world. Its major uses include:

- Agriculture - Control of aphids, cane beetles, thrips, stink bugs, locusts, and a variety of other insects that damage crops

- Arboriculture - Control of the emerald ash borer, hemlock woolly adelgid, and other insects that attack trees (including hemlock, maple, oak, and birch)

- Home Protection - Control of termites, carpenter ants, cockroaches, and moisture-loving insects

- Domestic animals - Control of fleas (applied to the neck)

- Turf - Control of Japanese beetle larvae

- Gardening - Control of aphids and other pests

When used on plants, imidacloprid, which is systemic, is slowly taken up by plant roots and slowly translocated up the plant via xylem tissue.

Application to Trees

When used on trees, it can take 30–60 days to reach the top (depending on the size and height) and enter the leaves in high enough quantities to be effective. Imidacloprid can be found in the trunk, the branches, the twigs, the leaves, the leaflets, and the seeds. Many trees are wind pollinated. But others such as fruit trees, linden, catalpa, and black locust trees are bee and wind pollinated and imidacloprid would likely be found in the flowers in small quantities. Higher doses must be used to control boring insects than other types.

Background

On January 21, 1986 a patent was filed, and granted on May 3, 1988, for imidacloprid in the United States (U.S. Pat. No. 4,742,060) by Nihon Tokushu Noyaku Seizo K.K. of Tokyo, Japan.

On March 25, 1992, Miles, Inc. (later Bayer CropScience) applied for registration of imidacloprid for turfgrass and ornamentals in the United States. On March 10, 1994, the U.S. Environmental Protection Agency approved the registration of imidacloprid.

On January 26, 2005, the Federal Register notes the establishment of the '(Pesticide Tolerances for) Emergency Exemptions' for imidacloprid. It use was granted to *Hawaii (for the) use (of) this pesticide on bananas(,) and the States of Minnesota, Nebraska, and North Dakota to use (of) this pesticide on sunflower(s).*

Brand Names

Imidacloprid has many brands and formulations for a wide range of uses, from delousing or de-fleaing animals to protecting trees. Selected brand names include Admire, Advantage (Advocate) (flea killer for pets), Centerfire 75, Confidor, Conguard, Gaucho, Hachikusan, Intercept, InVict, Kohinor, Mallet, Maxforce Quantum, Merit, Nuprid, Optrol, Premise, Prothor, Provado, Turfthor, Temprid (Bayer), Winner, and Xytect.

Biochemistry

Imidacloprid is a systemic chloronicotinyl pesticide, belonging to the class of neonicotinoid insecticides. It works by interfering with the transmission of nerve impulses in insects by binding irreversibly to specific insect nicotinic acetylcholine receptors.

As a systemic pesticide, imidacloprid translocates or moves easily in the xylem of plants from the soil into the leaves, fruit, pollen, and nectar of a plant. Imidacloprid also exhibits excellent translaminar movement in plants and can penetrate the leaf cuticle and move readily into leaf tissue.

Since imidacloprid is efficacious at very low levels (nanogram and picogram), it can be applied at lower concentrations (e.g., 0.05–0.125 lb/acre or 55–140 g/ha) than other insecticides. The availability of imidacloprid and its favorable toxicity package as compared to other insecticides on the market in the 1990s allowed the EPA to replace more toxic insecticides including the acetylcholinesterase inhibitors, the organophosphorus compounds, and methylcarbamates.

Environmental Fate

The main routes of dissipation of imidacloprid in the environment are aqueous photolysis (half-life = 1–4 hours) and plant uptake. The major photometabolites include imidacloprid desnitro, imidacloprid olefine, imidacloprid urea, and five minor metabolites. The end product of photodegradation is chloronicotinic acid (CNA) and ultimately carbon dioxide. Since imidacloprid has a low vapor pressure, it normally does not volatilize readily.

Although imidacloprid breaks down rapidly in water in the presence of light, it remains persistent in water in the absence of light. It has a water solubility of .61 g/L, which is relatively high. In the dark, at pH between 5 and 7, it breaks down very slowly, and at pH 9, the half-life is about 1 year. In soil under aerobic conditions, imidacloprid is persistent with a half-life of the order of 1–3 years. On the soil surface the half-life is 39 days. Major soil metabolites include imidacloprid nitrosimine, imidacloprid desnitro and imidacloprid urea, which ultimately degrade to 6-chloronicotinic acid, CO_2, and bound residues. 6-Chloronicotinic acid is recently shown to be mineralized via a nicotinic acid (vitamin B3) pathway in a soil bacterium.

In soil, imidacloprid strongly binds to organic matter. When not exposed to light, imidacloprid breaks down slowly in water, and thus has the potential to persist in groundwater for extended periods. However, in a survey of groundwater in areas of the United States which had been treated with imidacloprid for the emerald ash borer, imidacloprid was usually not detected. When detected, it was present at very low levels, mostly at concentrations less than 1 part per billion (ppb) with a maximum of 7 ppb, which are below levels of concern for human health. The detections have generally occurred in areas with porous rocky or sandy soils with little organic matter, where the risk of leaching is high — and/or where the water table was close to the surface.

Based on its high water solubility (0.5-0.6 g/L) and persistence, both the U.S. Environmental Protection Agency and the Pest Management Regulatory Agency in Canada consider imidacloprid to have a high potential to run off into surface water and to leach into ground water and thus warn not to apply it in areas where soils are permeable, particularly where the water table is shallow.

According to standards set by the environmental ministry of Canada, if used correctly (at recommended rates, without irrigation, and when heavy rainfall is not predicted), imidacloprid does not characteristically leach into the deeper soil layers despite its high water solubility (Rouchaud et al. 1994; Tomlin 2000; Krohn and Hellpointner 2002). In a series of field trials conducted by Rouchaud et al. (1994, 1996), in which imidacloprid was applied to sugar beet plots, it was consistently demonstrated that no detectable leaching of imidacloprid to the 10–20 cm soil layer occurred. Imidacloprid was applied to a corn field in Minnesota, and no imidacloprid residues were found in sample column segments below the 0-15.2 cm depth segment (Rice et al. 1991, as reviewed in Mulye 1995).

However, a 2012 water monitoring study by the state of California, performed by collecting agricultural runoff during the growing seasons of 2010 and 2011, found imidacloprid in 89% of samples, with levels ranging from 0.1-3.2 µg/L. 19% of the samples exceeded the EPA threshold for chronic toxicity for aquatic invertebrates of 1.05 µg/L. The authors also point out that Canadian and European guidelines are much lower (0.23 µg/L and 0.067 µg/L, respectively) and were exceeded in 73% and 88% of the samples, respectively. The authors concluded that "imidacloprid commonly moves offsite and contaminates surface waters at concentrations that could harm aquatic invertebrates".

Toxicology

Based on laboratory rat studies, imidacloprid is rated as "moderately toxic" on an acute oral basis to mammals and low toxicity on a dermal basis by the World Health Organization and the United States Environmental Protection Agency (class II or III, requiring a "Warning" or "Caution" label). It is rated as an "unlikely" carcinogen and as weakly mutagenic by the U.S. EPA (group E). It is not listed for reproductive or developmental toxicity, but is listed on EPA's Tier 1 Screening Order for chemicals to be tested under the Endocrine Disruptor Screening Program (EDSP). Tolerances for imidacloprid residues in food range from 0.02 mg/kg in eggs to 3.0 mg/kg in hops.

Animal toxicity is moderate when ingested orally and low when applied dermally. It is not irritating to eyes or skin in rabbits and guinea pigs (although some commercial preparations contain clay as an inert ingredient, which may be an irritant). The acute inhalation LD_{50} in rats was not reached at the greatest attainable concentrations, 69 milligrams per cubic meter of air as an aerosol, and 5,323 mg a.i./m³ of air as a dust. In rats subjected to a two-year feeding study, no observable effect was seen at 100 parts per million (ppm). In rats, the thyroid is the organ most affected by imidacloprid. Thyroid lesions occurred in male rats at a LOAEL of 16.9 mg a.i./kg/day. In a one-year feeding study in dogs, no observable effect was seen at 1,250 ppm, while levels up to 2,500 ppm led to hypercholesterolemia and elevated liver cytochrome p-450 measurements.

Bees and Other Insects

To members of the genus *Apis*, the honey bees, imidacloprid is one of the most toxic chemicals ever created as an insecticide. The acute oral LD_{50} ranges from 5 to 70 picograms of active ingredient per bee, making it more toxic to bees than the organophosphate dimethoate (oral LD_{50} 0.152 µg/bee) or the pyrethroid cypermethrin (oral LD_{50} 0.160 µg/bee). (For comparison, the weight of just the DNA of a human cell is about 7 picograms.) The toxicity of imidacloprid to bees differs from most insecticides in that it is more toxic orally than by contact. The contact acute LD_{50} is 0.024 µg active ingredient per bee.

Imidacloprid was first widely used in the United States in 1996 as it replaced three broad classes of insecticides. In 2006, U.S. commercial migratory beekeepers reported sharp declines in their honey bee colonies. Such declines had happened in the past; however unlike the case in previous losses, adult bees were abandoning their hives. Scientists named this phenomenon colony collapse disorder (CCD). Reports show that beekeepers in most states have been affected by CCD. Although no single factor has been identified as causing CCD, the United States Department of Agriculture (USDA) in their progress report on CCD stated that CCD may be "a syndrome caused by many different factors, working in combination or synergistically." Several studies have found that sub-lethal levels of imidacloprid increase honey bee susceptibility to the pathogen *Nosema*.

Dave Goulson (2012) of the University of Stirling showed that trivial effects of imidacloprid in lab and greenhouse experiments can translate into large effects in the field. The research found that bees consuming the pesticide suffered an 85% loss in the number of queens their hives produced, and a doubling of the number of bees who failed to return from food foraging trips.

Lu et al. (2012) reported they were able to replicate CCD with sub-lethal doses of imidacloprid. The imidacloprid-treated hives were nearly empty, consistent with CCD, and the authors exclude *Varroa* or *Nosema* as contributing causes.

In May 2012, researchers at the University of San Diego released a study showing that honey bees treated with a small dose of imidacloprid, comparable to what they would receive in nectar and formerly considered a safe amount, became "picky eaters," refusing nectars of lower sweetness and preferring to feed only on sweeter nectar. It was also found that bees exposed to imidacloprid performed the "waggle dance," the movements that bees use to inform hive mates of the location of foraging plants, at a lower rate.

Researchers from the Canadian Forest Service showed that imidacloprid used on trees at realistic field concentrations decreases leaf litter breakdown owing to adverse sublethal effects on non-target terrestrial invertebrates. The study did not find significant indication that the invertebrates, which normally decompose leaf litter, preferred uncontaminated leaves, and concluded that the invertebrates could not detect the imidacloprid.

A 2012 *in situ* study provided strong evidence that exposure to sublethal levels of imidacloprid in high fructose corn syrup (HFCS) used to feed honey bees when forage is not available causes bees to exhibit symptoms consistent to CCD 23 weeks post imidacloprid dosing. The researchers suggested that "the observed delayed mortality in honey bees caused by imidacloprid in HFCS is a novel and plausible mechanism for CCD, and should be validated in future studies".

Sublethal doses (<10 ppb) to aphids have been found to lead to altered behavior, such as wandering and eventual starvation. Very low concentrations also reduced nymph viability. In bumblebees exposure to 10 ppb imidacloprid reduces natural foraging behaviour, increases worker mortality and leads to reduced brood development. A 2013 study showed that bumblebee colonies exposed to 10 ppb of imidacloprid started failing after three weeks when the death rate increased and the birth rate decreased. The researchers attributed this to exposed colonies performing essential tasks, such as foraging, thermoregulation and brood care, less well than unexposed colonies. This suggests that sublethal imidacloprid causes colony failure through reduced colony function.

In January 2013, the European Food Safety Authority stated that neonicotinoids pose an unacceptably high risk to bees, and that the industry-sponsored science upon which regulatory agencies' claims of safety have relied might be flawed, concluding that, "A high acute risk to honey bees was identified from exposure via dust drift for the seed treatment uses in maize, oilseed rape and cereals. A high acute risk was also identified from exposure via residues in nectar and/or pollen." An author of a *Science* study prompting the EFSA review suggested that industry science pertaining to neonicotinoids may have been deliberately deceptive, and the UK Parliament has asked the manufacturer Bayer Crop Science to explain discrepancies in evidence they have submitted to an investigation.

Birds

In bobwhite quail (*Colinus virginianus*), imidacloprid was determined to be moderately toxic with an acute oral LD$_{50}$ of 152 mg a.i./kg. It was slightly toxic in a 5-day dietary study with an acute oral LC$_{50}$ of 1,420 mg a.i./kg diet, a NOAEC of < 69 mg a.i./kg diet, and a LOAEC = 69 mg a.i./kg diet. Exposed birds exhibited ataxia, wing drop, opisthotonos, immobility, hyperactivity, fluid-filled crops and intestines, and discolored livers. In a reproductive toxicity study with bobwhite quail, the NOAEC = 120 mg a.i./kg diet and the LOAEC = 240 mg a.i./kg diet. Eggshell thinning and decreased adult weight were observed at 240 mg a.i./kg diet.

Imidacloprid is highly toxic to four bird species: Japanese quail, house sparrow, canary, and pigeon. The acute oral LD$_{50}$ for Japanese quail (*Coturnix coturnix*) is 31 mg a.i./kg bw with a NOAEL = 3.1 mg a.i./kg. The acute oral LD$_{50}$ for house sparrow (*Passer domesticus*) is 41 mg a.i./kg bw with a NOAEL = 3 mg a.i./kg and a NOAEL = 6 mg a.i./kg. The LD$_{50}$s for pigeon (*Columba livia*) and canary (*Serinus canaria*) are 25–50 mg a.i./kg. Mallard ducks are more resistant to the effects of imidacloprid with a 5-day dietary LC$_{50}$ of > 4,797 ppm. The NOAEC for body weight and feed consumption is 69 mg a.i./kg diet. Reproductive studies with mallard ducks showed eggshell thinning at 240 mg a.i./kg diet. According to the European Food Safety Authority, imidacloprid poses a potential high acute risk for herbivorous and insectivorous birds and granivorous mammals.

Chronic risk has not been well established. The hypothesis that imidacloprid has a negative impact on insectivorous bird populations is supported by a study of bird population trends in the Netherlands, where correlation has been identified between surface-water concentrations of imidacloprid and population decline. At imidacloprid concentrations of more than 20 nanograms per litre, bird populations tended to decline by 3.5 per cent on average annually. Additional analyses in this study revealed that spatial pattern of bird population decline appeared only after the introduction of imidacloprid to the Netherlands, in the mid-1990s, and that this correlation is not linked to any other land usage factor.

Aquatic Life

Imidacloprid is highly toxic on an acute basis to aquatic invertebrates, with EC_{50} values = 0.037 - 0.115 ppm. It is also highly toxic to aquatic invertebrates on a chronic basis (effects on growth and movement): NOAEC/LOAEC = 1.8/3.6 ppm in daphnids; NOAEC = 0.001 in *Chironomus* midge, and NOAEC/LOAEC = 0.00006/0.0013 ppm in mysid shrimp. Its toxicity to fish is relatively low; however, the EPA has requested review of secondary effects on fish with food chains that include sensitive aquatic invertebrates.

Plant Life

Imidacloprid has been shown to turn off some genes that some rice varieties use to produce defensive chemicals. While imidacloprid is used for control of the brown planthopper and other rice pests, there is evidence that imidacloprid actually increases the susceptibility of the rice plant to planthopper infestation and attacks.

Health Impact

Imidacloprid and its nitrosoimine metabolite (WAK 3839) have been well studied in rats, mice and dogs. In mammals, the primary effects following acute high-dose oral exposure to imidacloprid are mortality, transient cholinergic effects (dizziness, apathy, locomotor effects, labored breathing) and transient growth retardation. Exposure to high doses may be associated with degenerative changes in the testes, thymus, bone marrow and pancreas. Cardiovascular and hematological effects have also been observed at higher doses. The primary effects of longer term, lower-dose exposure to imidacloprid are on the liver, thyroid, and body weight (reduction). Low- to mid-dose oral exposures have been associated with reproductive toxicity, developmental retardation and neurobehavioral deficits in rats and rabbits. Imidacloprid is neither carcinogenic in laboratory animals nor mutagenic in standard laboratory assays.

No studies have been published involving human subjects chronically exposed to imidacloprid. Effects of imidacloprid on human health and the environment depend on how much imidacloprid is present and the length and frequency of exposure. Effects also depend on the health of a person and/or certain environmental factors.

A study conducted in tissue culture of neurons harvested from newborn rats showed that Imidacloprid and acetamiprid, another neonicotinoid, excited the neurons in a way similar to nicotine, so the effects of neonicotinoids on developing mammalian brains might be similar to the adverse effects of nicotine.

Overdosage

Persons who might orally ingest acute amounts would experience emesis, diaphoresis, drowsiness and disorientation. This would need to be intentional since a large amount would need to be ingested to experience a toxic reaction. In dogs the LD_{50} is 450 mg/kg of body weight (i.e., in any sample of medium sized dogs weighing, say, 13 kilograms (29 lb), half of them would be killed after consuming 5,850 mg of imidacloprid, or about $\frac{1}{5}$th of an ounce) . Blood imidacloprid concentrations may be measured to confirm diagnosis in hospitalized patients or to establish the cause of death in postmortem investigations.

Clothianidin

Clothianidin is an insecticide developed by Takeda Chemical Industries and Bayer AG. Similar to thiamethoxam and imidacloprid, it is a neonicotinoid. Neonicotinoids are a class of insecticides that are chemically similar to nicotine, which has been used as a pesticide since the late 1700s. Clothianidin and other neonicotinoids act on the central nervous system of insects as an agonist of acetylcholine, the neurotransmitter that stimulates nAChR, targeting the same receptor site (AChR) and activating post-synaptic acetylcholine receptors but not inhibiting AChE. Clothianidin and other neonicotinoids were developed to last longer than nicotine, which is more toxic and which breaks down too quickly in the environment. However, studies published in 2012 show that neonicotinoid dust released at planting time may persist in nearby fields for several years and be taken up into non-target plants, which are then foraged by bees and other insects.

Clothianidin is an alternative to organophosphate, carbamate, and pyrethroid pesticides. It poses lower risks to mammals, including humans, when compared to organophosphates and carbamates. It has helped prevent insect pests build up resistance to organophosphate and pyrethroid pesticides.

According to the Environmental Protection Agency (EPA), clothianidin's major risk concern is to nontarget insects (honey bees). Information from standard tests and field studies, as well as incident reports involving other neonicotinoid insecticides (e.g., imidacloprid) suggest the potential for long term toxic risk to honey bees and other beneficial insects. In January 2013, the European Food Safety Authority stated that neonicotinoids including clothianidin pose an unacceptably high risk to bees, concluding, "A high acute risk to honey bees was identified from exposure via dust drift for the seed treatment uses in maize, oilseed rape and cereals. A high acute risk was also identified from exposure via residues in nectar and/or pollen."

Authorized Uses

Clothianidin is authorized for spray, dust, soil drench (for uptake via plant roots), injectable liquid (into tree limbs and trunks, sugar cane stalks etc.), and seed treatment uses, in which clothianidin coats seeds that take up the pesticide via the roots as the plant grows. The chemical may be used to protect plants against a wide variety of agricultural pests in many countries, of which the following are mentioned in citable English-language sources: Australia, Austria, Belgium, Bulgaria, Canada, Czech Republic, Denmark, Estonia, France, Finland, Germany, Greece, Hungary, Italy, Ireland,

Japan, Korea, Lithuania, Netherlands, New Zealand, Poland, Portugal, Serbia, Slovakia, Slovenia, Spain, UK, and the United States. Seed treatment uses of clothianidin, corn in particular, have been revoked or suspended in Germany, Italy and Slovenia. The suspensions are reflective of E.U. pesticide law and are generally associated with acute poisoning of bees from pesticide dust being blown off of treated seeds, especially corn, and onto nearby farms where bees were performing pollinator services.

Background

Although nicotine has been used as a pesticide for over 200 years it degraded too rapidly in the environment and lacked the selectivity to be very useful in large-scale agricultural situations. However, in order to address this problem, the neonicotinoids (chloronicotinyl insecticides) were developed as a substitute of nicotine. Clothianidin is an alternative to organophosphate, carbamate, and pyrethroid pesticides. It poses lower risks to mammals, including humans, when compared to organophosphates and carbamates. It also plays a key role helping to prevent the buildup in insect pests of resistance to organophosphate and pyrethroid pesticides, which is a growing problem in parts of Europe.

Clothianidin was first given conditional registration for use as a pesticide by the United States Environmental Protection Agency in 2003, pending the completion of additional study of its safety to be done by December 2004. Bayer did not complete the study on time and asked for an extension. The date was postponed to May 2005 and they also granted Bayer the permission it had sought to conduct its study on canola in Canada, instead of on corn in the United States. The study was not completed until 2007. In a November 2007 memo EPA scientists declared the study "scientifically sound," adding that it, "satisfies the guideline requirements for a field toxicity test with honeybees."

Clothianidin continued to be sold under a conditional registration, and in April 2010 it was granted an unconditional registration for use as a seed treatment for corn and canola. However, in response to concerns raised by bee keepers, in November the EPA released a memorandum in which they stated that some of the studies submitted did not appear to be adequate and the unconditional registration was withdrawn.

In 2012, arguing that after more than 9 years the EPA continues to maintain the registration status for clothianidin despite the fact that the registrant has failed to supply satisfactory studies confirming its safety, an alliance of beekeepers and environmental groups filed a petition on March 21 asking the EPA to block the use of clothianidin in agricultural fields until they have conducted a review of the product. The petitioners state that they are aware that the EPA has moved up its registration review of clothianidin and other neonicotinoids in response to concerns about their impacts on pollinators, however they note that this process is projected by the EPA to take six to eight years and is thus grossly insufficient to address the urgency of the threat to pollinators.

Toxicity

Regulatory authorities describe the toxicological database for clothianidin as "extensive", and many studies have been reviewed to support registrations around the globe for this chemical. Laboratory and field testing revealed that clothianidin shows relatively low toxicity to many test species but is highly or very highly toxic to others. Toxicity varies depending on whether the exposure occurs on a short-term (acute) or long-term (chronic) basis.

Because it is systemic, persistent and highly toxic to honey bees, the Pest Management Regulatory Agency of Canada has requested additional data to fully assess the potential effects of chronic exposure of clothianidin, resulting from its potential movement into plant pollen and nectar.

Mammals

Clothianidin is moderately toxic in the short-term to mammals that eat it, and long-term ingestion may result in reproductive and/or developmental effects. Using laboratory test animals as surrogates for humans and dosages much higher than are expected from exposure related to actual use, rats showed low short-term oral, dermal, and inhalation toxicity to clothianidin. For mice, acute oral toxicity was moderate to high. Rabbits showed little to no skin or eye irritation when exposed to clothianidin, and the skin of guinea pigs was not sensitized by it. When extrapolated to humans, these results suggest that clothianidin is moderately toxic through oral exposure, but toxicity is low through skin contact or inhalation. While clothianidin may cause slight eye irritation, it is not expected to be a skin sensitizer or irritant. Clothianidin does not damage genetic material nor is there evidence that it causes cancer in rats or mice; it is unlikely to be a human carcinogen.

Permissible amounts of clothianidin residue on food and animal feed vary from crop to crop and nation to nation. However, regulatory authorities around the globe emphasize that when used according to the label instructions, clothianidin residues on food are not expected to exceed safe levels (as defined by each nation's laws and regulations).

Aquatic Life

In the 2003 United States EPA assessment report it was stated that clothianidin should not present a direct acute or chronic risk to freshwater and estuarine/marine fish, or a risk to terrestrial or aquatic vascular and nonvascular plants. It is considered to be toxic to aquatic invertebrates if disposal of wastes according to disposal instructions are not followed. The Pest Management Regulatory Agency of Canada lists it as "very highly toxic" to aquatic invertebrates, but only slightly toxic to fish.

In the 2003 EPA report it was stated that although no water monitoring studies had been conducted, due to the extreme mobility and persistence of clothianidin in the environment, clothianidin has the properties of a chemical which could lead to widespread groundwater contamination should the registrant (e.g. Bayer or Takeda) request field uses involving direct application of clothianidin to the land surface. In a 2010 EPA report, it was noted that the registrant (e.g. Bayer or Takeda) had recently added new uses on the labels, including using the pesticide directly applied to the soil surface/foliage at much higher application rate than as specified in 2003. As a result, the potential for clothianidin to move from the treated area to the nearby surface water body under the new uses is much greater than with use as a seed treatment.

Birds

According to the EPA, clothianidin is practically non-toxic to test bird species that were fed relatively large doses of the chemical on an acute basis. However, EPA assessments show that exposure to treated seeds through ingestion may result in chronic toxic risk to non-endangered and endangered small birds (e.g., songbirds). Bobwhite quail eggshell thickness was affected when the test birds were given a diet consisting of relatively large amounts of clothianidin-treated seeds. The Pest Management Regulatory Agency of Canada lists clothianidin as "moderately toxic" to birds.

Bees and Other Insect Pollinators

Honey bees pollinate crops responsible for about a third of the human diet; about $224 billion worth of crops worldwide. Beginning in 2006, beekeepers in the United States began to report unexplained losses of hives — 30 percent and upward — leading to a phenomenon called colony collapse disorder (CCD). The cause of CCD remains under debate, but scientific consensus is beginning to emerge suggesting that there is no one cause but rather a combination of factors including lack of foraging plants, infections, breeding, and pesticides—with none catastrophic on their own, but having a synergistic effect when occurring in combination.

The Australian Pesticides and Veterinary Medicines Authority notes that clothianidin ranks "among the most highly acutely toxic insecticides to bees" through contact and oral exposure. Since clothianidin is a systemic pesticide that is taken up by the plant, there is also potential for toxic chronic exposure resulting in long-term effects to bees and other pollinators from clothianidin residue in pollen and nectar. According to the Environmental Protection Agency (EPA), in addition to potential effects on worker bees, there are also concerns about lethal and/or sub-lethal effects in the larvae and reproductive effects in the queen from chronic exposure. However, in a 2012 statement the EPA reported that they are not aware of any data demonstrating that bee colonies are subject to elevated losses due to long-term exposure when clothianidin products are used at authorized rates.

Honey bees and other pollinators are particularly sensitive to clothianidin, as evidenced by the results of laboratory and field toxicity testing and demonstrated in acute poisoning incidents in France and Germany in 2008, and in Canada in 2010 and 2013 associated with the planting of corn seeds treated with clothianidin. To reduce the risk to pollinators from acute exposure to clothianidin sprays, label instructions prohibit the use of these products when crops or weeds are in bloom and pollinators are nearby, but in the U.S. label instructions do not require the use of a "sticker", a sticking agent meant to reduce dust from treated seeds during planting. However, according to the EPA, the use of sticking agents to reduce dust from treated seeds is standard practice in the U.S.

In a July 2008 German beekill incident, German beekeepers reported that 50 to 100 percent of their hives had been lost after pneumatic equipment used to plant corn seed blew clouds of pesticide dust into the air, which was then pushed by the wind onto neighboring canola fields in which managed bees were performing pollinator services. The accident was found to be the result of improper planting procedures and the weather. However, in 2009, Germany suspended authorization for the use of clothianidin on corn, citing unanswered questions that remained about potential exposure of bees and other pollinators to neonicotinoid pesticides.

A 2011 Congressional Research Report describing some of the reasons why scientists believe honey bee colonies are being affected by CCD reported that the United States Department of Agriculture had concluded in 2009, "it now seems clear that no single factor alone is responsible for the malady." According to the research report, the neonicotinoids, which contain the active ingredient imidacloprid, and similar other chemicals, such as clothianidin and thiamethoxam, are being studied for a possible link to CCD. Honey bees are thought to possibly be affected by such chemicals, which are known to work their way through the plant up into the flowers and leave residues in the nectar and pollen that bees forage on. The scientists studying CCD have tested samples of pollen and have indicated findings of a broad range of substances, including insecticides, fungicides, and herbicides. They note that the doses taken up by bees are not lethal, but they are concerned about possible chronic problems caused by long-term exposure.

A report released in 2012 found a close relationship between the deaths of bees and the use of pneumatic drilling machines for the sowing of corn seeds coated with clothianidin and other neonicotinoid insecticides. In pneumatic drilling machines, seeds are sucked in, causing the erosion of fragments of the insecticide shell, which are then expelled with a current of air. Field tests found that foraging bees flying through dust released during the planting of corn seeds coated with neonicotinoid insecticides may encounter exposure high enough to be lethal. They concluded: "The consequent acute lethal effect evidenced in all the field sowing experiments can be well compared with the colony loss phenomena widely reported by beekeepers in spring and often associated to corn sowing." Another field study released in 2012 looked at sublethal effects of clothianidin and imidacloprid in amounts that bees might be exposed to during foraging. Sublethal doses can affect orientation, foraging, learning ability and brood care. The study found: "clothianidin elicited detrimental sub-lethal effects at somewhat lower doses (0.5 ng/bee) than imidacloprid (1.5 ng/bee). Bees disappeared at the level of 1 ng for clothianidin, while we could register the first bee losses for imidacloprid at doses exceeding 3 ng."

In a 2012 study, scientists found that an analyses of bees found dead in and around hives from several apiaries in Indiana showed the presence of the neonicotinoid insecticides clothianidin and thiamethoxam. The research showed that the insecticides were present at high concentrations in waste talc that was exhausted from farm machinery during planting and that is left outside after cleaning the planting equipment. Talc is used in the vacuum system planters to keep pesticide treated seeds flowing freely and was studied by the investigators since the waste talc can be picked up by the wind, and could spread the pesticide to non-treated areas; they did not however investigate whether and how much pesticide spreads this way. The insecticides were also consistently found at low levels in soil up to two years after treated seed was planted, and on nearby dandelion flowers and corn pollen gathered by the bees. Also in 2012, researchers in Italy published findings that the pneumatic drilling machines that plant corn seeds coated with clothianidin and imidacloprid release large amounts of the pesticide into the air, causing significant mortality in foraging honey bees.

Neonicotinoids Banned by European Union

In 2012, several peer reviewed independent studies were published showing that neonicotinoids, including clothianidin, had previously undetected routes of exposure affecting bees including through dust, pollen, and nectar; that sub-nanogram toxicity resulted in failure to return to the hive without immediate lethality, the primary symptom of colony collapse disorder; and showing environmental persistence in agricultural irrigation channels and soil. These reports prompted a formal peer review by the European Food Safety Authority (EFSA), which stated in January 2013 that neonicotinoids, including clothianidin, pose an unacceptably high risk to bees, and that the industry-sponsored science upon which regulatory agencies' claims of safety have relied on may be flawed and contain several data gaps not previously considered. Their review concluded, "A high acute risk to honey bees was identified from exposure via dust drift for the seed treatment uses in maize, oilseed rape and cereals. A high acute risk was also identified from exposure via residues in nectar and/or pollen." In April 2013, the European Union voted for a two-year restriction on neonicotinoid insecticides. The ban restricted the use of imidacloprid, clothianidin, and thiamethoxam for use on crops that are attractive to bees (maize, cotton, sunflower, and rapeseed), and goes into effect on December 1, 2013. Eight nations voted against the motion, including the British government which argued that the science was incomplete.

Following on the release of the EFSA report in January 2013, the UK Parliament has asked manufacturer Bayer Cropscience to explain discrepancies in evidence they have submitted to an investigation.

Data Gaps

North American and European pesticide regulatory authorities have identified specific data gaps and uncertainties for which clothianidin manufacturers must provide data. Studies required of the manufacturers will further investigate clothianidin's:

- environmental persistence in soil and subsequent uptake in rotational crops

- availability in pollen and nectar

- long-term effects on honey bees and other pollinators

- developmental immunotoxicity

- effects on aerobic aquatic metabolism

- ability to leach from treated seeds and

- acute toxicity to freshwater invertebrates

The challenges associated with studying potential long-term effects of pesticides on honey bee colonies are well documented and include the inability to adequately monitor individual bee health or extrapolate effects on individuals to whole hives. Behavior changes between bees and/or colonies in laboratory or field test conditions versus natural environments also add to the challenges. Studies submitted by Bayer AG to USEPA have provided some useful information about clothianidin's potential long-term effects on honey bees but outstanding questions remain. USEPA's analysis of nine pollinator field studies submitted concluded that three were invalid, so EPA did not use the data they provided in making its regulatory decision for clothianidin. EPA classified the remainder as supplemental, generally because Bayer AG conducted the studies without EPA first approving the protocols. Supplemental studies are ones that don't definitively answer uncertainties but still provide some data that might be useful in characterizing risk. Indicative of the rapid advance of regulators' understanding of pollinator science, USEPA first accepted one of the studies as sound science in 2007, then reclassified it as invalid in November 2010 only to reclassify it as supplemental one month later. The changes in EPA's classification of this study have no effect on the regulatory status for clothianidin in the U.S. because the study does not provide data with which EPA can legally justify altering its 2003 registration decision. An international group of pesticide regulators, researchers, industry representatives, and beekeepers is working to develop a study protocol that will definitively answer remaining questions about the potential long-term effects on bee colonies and other pollinators.

Environmental Persistence

Laboratory and field testing shows that clothianidin is persistent and mobile in the environment, stable to hydrolysis, and has potential to leach to ground water and be transported via runoff to surface water bodies. Worst-case scenario estimates indicate that if applied at the maximum rate repeatedly over years, clothianidin has the potential to accumulate in the top 15 cm of soil. However, the Australian pesticide authority's review of rotational crop studies determined that clothianidin generally is not taken up by crops sown in fields where treated corn seeds were planted, even when the test corn seeds were coated with an intentionally large amount of the chemical (2 mg/seed vs the authorized maximum application rate of 1.25 mg).

Risk Mitigation

Once laboratory and field data identify hazards associated with a chemical, regulatory authorities take different approaches to mitigate those hazards and bring the risks down to acceptable levels, as defined by each nation's laws and regulations. For clothianidin, hazard mitigation includes establishing the maximum amount of the chemical that can be used (e.g. kg/acre or mg/seed), requiring buffer zones around treated fields to protect water supplies, and prohibiting the use of low-technology seed treatment methods or equipment that can send clouds of clothianidin dust or spray up into the air during seeding operations.

Clothianidin users are also required to monitor the weather and not use the chemical or seeds treated with it on windy days or when rain is forecast. Workers are protected from clothianidin exposure through requirements for personal protective equipment, such as long-sleeve shirts, gloves, long pants, boots, and face mask or respirators as appropriate. To reduce the possibility that birds and small mammals might eat treated seeds, users are required to ensure that soil covers planted seeds and that any spilled seed is picked up.

Chlorpyrifos

Chlorpyrifos (IUPAC name: *O,O*-diethyl *O*-3,5,6-trichloropyridin-2-yl phosphorothioate) is a crystalline organophosphate insecticide, acaracide and miticide. It was introduced in 1965 by Dow Chemical Company and is known by many trade names, including Dursban and Lorsban. It acts on the nervous system of insects by inhibiting acetylcholinesterase.

Chlorpyrifos is moderately toxic to humans, and exposure has been linked to neurological effects, persistent developmental disorders and autoimmune disorders. Exposure during pregnancy retards the mental development of children, and most home use was banned in 2001 in the U.S. In agriculture, it is "one of the most widely used organophosphate insecticides" in the United States, according to the United States Environmental Protection Agency (EPA), and before being phased out for residential use was one of the most used residential insecticides.

Manufacture

Chlorpyrifos is produced via a multistep synthesis from 3-methylpyridine, eventually reacting 3,5,6-trichloro-2-pyridinol with diethylthiophosphoryl chloride.

Uses

Chlorpyrifos is used around the world to control pest insects in agricultural, residential and commercial settings. Its use in residential applications is restricted in multiple countries. According to Dow, chlorpyrifos is registered for use in nearly 100 countries and is annually applied to approximately 8.5 million crop acres. The crops with the most use are cotton, corn, almonds and fruit trees including oranges, bananas and apples.

Chlorpyrifos was first registered for use in the United States in 1965 for control of foliage and soil-born insects. The chemical became widely used in residential settings, on golf course turf, as a structural termite control agent, and in agricultural use. Most residential use has been phased out in the United States; however it remains a common agricultural insecticide.

EPA estimated that between 1987 and 1998 about 21 million pounds of chlorpyrifos were annually used in the US. In 2007, chlorpyrifos was the most commonly used organophosphate pesticide in the United States, with an estimated 8 to 11 million pounds applied, and the 14th most common agricultural pesticide ingredient overall in 2007 in the United States.

Application

Chlorpyrifos is normally supplied as a 23.5% or 50% liquid concentrate. The recommended concentration for direct-spray pin point application is 0.5% and for wide area application a 0.03 – 0.12% mix is recommended (US).

Toxicity and Safety

Chlorpyrifos exposure may lead to acute toxicity at higher doses. Persistent health effects follow acute poisoning or from long-term exposure to low doses. developmental effects appear in fetuses and children even at very small doses.

Persistent Health Effects

Gestation, Infancy and Childhood

Epidemiological and experimental animal studies suggest that infants and children are more susceptible than adults to effects from low dose exposure. The young have a decreased capacity to detoxify chlorpyrifos and its metabolites. This results in disruption in nervous system developmental processes, as observed in animal experiments.

Human studies: In multiple epidemiological studies, chlorpyrifos exposure during gestation or childhood has been linked with lower birth weight and neurological changes such as slower motor development and attention problems. Exposure to organophosphate pesticides in general has been increasingly associated with changes in children's cognitive, behavioral and motor performance.

Animal experiments: In experiments with rats, early, short-term low-dose exposure to chlorpyrifos resulted in lasting neurological changes, with larger effects on emotional processing and cognition than on motor skills. Such rats exhibited behaviors consistent with depression and reduced anxiety. In rats, low-level exposure during development has its greatest neurotoxic effects during the period in which sex differences in the brain develop. Exposure leads to reductions or reversals of normal gender differences. Exposure to low levels of chlorpyrifos early in rat life or as adults also affects metabolism and body weight. These rats show increased body weight as well as changes in liver function and chemical indicators similar to prediabetes, likely associated with changes to the cyclic AMP system.

Adulthood

Adults may develop lingering health effects following acute exposure or repeated low-dose exposure. Among agricultural workers, chlorpyrifos has been associated with slightly increased risk of wheeze, a whistling sound while breathing due to airway obstruction in the airways.

Among 50 farm pesticides studied, chlorpyrifos was associated with higher risks of lung cancer among frequent pesticide applicators than among infrequent or non-users. Pesticide applicators as a whole were found to have a 50% lower cancer risk than the general public, likely due to their nearly 50% lower smoking rate. However, chlorpyrifos applicators had a 15% lower cancer risk than the general public, which the study suggests indicates a link between chlorpyrifos application and lung cancer.

Acute Health Effects

For acute effects, the World Health Organization classifies chlorpyrifos as Class II: moderately toxic. The oral LD50 in experimental animals is 32 to 1000 mg/kg. The dermal LD50 in rats is greater than 2000 mg/kg and 1000 to 2000 mg/kg in rabbits. The 4-hour inhalation LC50 for chlorpyrifos in rats is greater than 200 mg/m3.

Symptoms of Acute Exposure

Acute poisoning results mainly from interference with the acetylcholine neurotransmission pathway, leading to a range of neuromuscular symptoms. Relatively mild poisoning can result in eye watering, increased saliva and sweating, nausea and headache. Intermediate exposure may lead to muscle spasms or weakness, vomiting or diarrhea and impaired vision. Symptoms of severe poisoning include seizures, unconsciousness, paralysis, and suffocation from lung failure.

Children are more likely to experience muscle weakness rather than twitching; excessive saliva rather than sweat or tears; seizures; and sleepiness or coma.

Frequency of Acute Exposure

Acute poisoning is probably most common in agricultural areas in Asia, where many small farmers are affected. Poisoning may be due to occupational or accidental exposure or intentional self-harm. Precise numbers of chlorpyrifos poisonings globally are not available. Pesticides are used in an estimated 200,000+ suicides annually. Organophosphates are thought to constitute two-thirds of ingested pesticides in rural Asia. Chlorpyrifos is among the commonly used pesticides used for self-harm.

In the US, the number of incidents of chlorpyrifos exposure reported to the US National Pesticide Information Center shrank sharply from over 200 in the year 2000 to less than 50 in 2003, following the residential ban.

Treatment

Poisoning is treated with atropine and simultaneously with oximes such as pralidoxime. Atropine blocks acetylcholine from binding with muscarinic receptors, which reduces the pesticide's impact. However, atropine does not affect acetylcholine at nicotinic receptors and thus is a partial treatment. Pralidoxime is intended to reactivate acetylcholinesterase, but the benefit of oxime treatment is questioned. A randomized controlled trial (RCT) supported the use of higher doses of pralidoxime rather than lower doses. A subsequent double-blind RCT, that treated patients who self-poisoned, found no benefit of pralidoxime, including specifically in chlorpyrifos patients.

Tourist Deaths

Chlorpyrifos poisoning was described by New Zealand scientists as the likely cause of death of several tourists in Chiang Mai, Thailand who developed myocarditis in 2011. Thai investigators came to no conclusion on the subject, but maintain that chlorpyrifos was not responsible and that the deaths were not linked.

Mechanisms of Toxicity

Acetylcholine Neurotransmission

Primarily, chlorpyrifos and other organophosphate pesticides interfere with signaling from the neurotransmitter acetylcholine. One chlorpyrifos metabolite, chlorpyrifos-oxon, binds permanently to the enzyme acetylcholinesterase, preventing this enzyme from deactivating acetylcholine in the synapse. By irreversibly inhibiting acetylcholinesterase, chlorpyrifos leads to a build-up of acetylcholine between neurons and a stronger, longer-lasting signal to the next neuron. Only when new molecules of acetylcholinesterase have been synthesized can normal function return. Acute symptoms of chlorpyrifos poisoning only occur when more than 70% of acetylcholinesterase molecules are inhibited. This mechanism is well established for acute chlorpyrifos poisoning and also some lower-dose health impacts. It is also the primary insecticidal mechanism.

Non-cholinesterase Mechanisms

Chlorpyrifos may affect other neurotransmitters, enzymes and cell signaling pathways, potentially at doses below those that substantially inhibit acetylcholinesterase. The extent of and mechanisms for these effects remain to be fully characterized. Laboratory experiments in rats and cell cultures suggest that exposure to low doses of chlorpyrifos may alter serotonin signaling and increase rat symptoms of depression; change the expression or activity of several serine hydrolase enzymes, including neuropathy target esterase and several endocannabinoid enzymes; affect components of the cyclic AMP system; and influence other chemical pathways.

Paraoxonase Activity

The enzyme paraoxonase 1 (PON1) detoxifies chlorpyrifos oxon, the more toxic metabolite of chlorpyrifos, via hydrolysis. In laboratory animals, additional PON1 protects against chlorpyrifos toxicity while individuals that do not produce PON1 are particularly susceptible. In humans, studies about the effect of PON1 activity on the toxicity of chlorpyrifos and other organophosphates are mixed, with modest yet inconclusive evidence that higher levels of PON1 activity may protect against chlorpyrifos exposure in adults; PON1 activity may be most likely to offer protection from low-level chronic doses. Human populations have genetic variation in the sequence of PON1 and its promoter region that may influence the effectiveness of PON1 at detoxifying chlorpyrifos oxon and the amount of PON1 available to do so. Some evidence indicates that children born to women with low PON1 may be particularly susceptible to chlorpyrifos exposure. Further, infants produce low levels of PON1 until six months to several years after birth, likely increasing the risk from chlorpyrifos exposure early in life.

Combined Exposures

Several studies have examined the effects of combined exposure to chlorpyrifos and other chemical agents, and these combined exposures can result in different effects during development. Female rats exposed first to dexamethasone, a treatment for premature labor, for three days in utero and then to low levels of chlorpyrifos for four days after birth experienced additional damage to the acetylcholine system upstream of the synapse that was not observed with either exposure alone. In both male and female rats, combined exposures to dexamethasone and chlorpyrifos decreased serotonin turnover in the synapse, for female rats with a greater-than-additive result. Rats that were co-exposed to dexamethasone and chlorpyrifos also exhibited complex behavioral differences from exposure to either chemical alone, including lessening or reversing normal sex differences in behavior. In the lab, in rats and neural cells co-exposed to both nicotine and chlorpyrifos, nicotine appears to protect against chlorpyrifos acetylcholinesterase inhibition and reduce its effects on neurodevelopment. In at least one study, nicotine appeared to enhance chlorpyrifos detoxification.

Human Exposure

In 2011, EPA estimated that, in the general US population, people consume 0.009 micrograms of chlorpyrifos per kilogram of their body weight per day directly from food residue. Children are estimated to consume a greater quantity of chlorpyrifos per unit of body weight from food residue, with toddlers the highest at 0.025 micrograms of chlorpyrifos per kilogram of their body weight per day. People may also ingest chlorprifos from drinking water or from residue in food handling establishments. The EPA's acceptable daily dose is 0.3 micrograms/kg/day.

Before residential use was restricted in the US, data from 1999-2000 in the national NHANES study detected the metabolite TCPy in 91% of human urine samples tested. In samples collected between 2007 and 2009 from families living in Northern California, TCPy was found in 98.7% of floor wipes tested and in 65% of urine samples tested. For both children and adults, the average concentrations of TCPy in urine were lower in the later study. A 2008 study found dramatic drops in the urinary levels of chlorpyrifos metabolites when children in the general population switched from conventional to organic diets.

Certain populations with higher likely exposure to chlorpyrifos, such as people who apply pesticides, work on farms, or live in agricultural communities, have been measured in the US to excrete TCPy in their urine that are 5 to 10 times greater than levels in the general population.

Air monitoring studies conducted by the California Air Resources Board (CARB) documented chlorpyrifos in the air of California communities. Analyses indicate that children living in areas of high chlorpyrifos use are often exposed to levels that exceed EPA dosages. Advocacy groups monitored air samples in Washington and Lindsay, CA, in 2006 with comparable results. Grower and pesticide industry groups argued that the air levels documented in these studies are not high enough to cause significant exposure or adverse effects, but a follow-up biomonitoring study in Lindsay showed that people there display above-normal chlorpyrifos levels.

Effects on Wildlife

Aquatic Life

Among freshwater aquatic organisms, crustaceans and insects appear to be more sensitive to acute exposure than are fish. Aquatic insects and animals appear to absorb chlorpyrifos directly from water rather than ingesting it with their diet or through sediment exposure.

Concentrated chlorpyrifos released into rivers killed insects, shrimp and fish. In Britain, the rivers Roding (1985), Ouse (2001), Wey (2002 & 2003), and Kennet (2013) all experienced insect, shrimp, and/or fish kills as a result of small releases of concentrated chlorpyrifos. The July 2013 release along the River Kennet poisoned insect life and shrimp along 15 km of the river, potentially from several teaspoonsful of concentrated chlorpyrifos washed down a drain.

Bees

Acute exposure to chlorpyrifos can be toxic to bees, with an oral LD50 of 360 ng/bee and a contact LD50 of 70 ng/bee. Guidelines for Washington state recommend that chlorpyrifos products should not be applied to flowering plants such as fruit trees within 4–6 days of blossoming to prevent bees from directly contacting the residue.

Risk assessments have primarily considered acute exposure, but more recently researchers have begun to investigate the effects of chronic, low-level exposure through residue in pollen and components of bee hives. A review of US studies, several European countries, Brazil and India found chlorpyrifos in nearly 15% of hive pollen samples and just over 20% of honey samples. Because of its high toxicity and prevalence in pollen and honey, bees are considered to have higher risk from chlorpyrifos exposure via their diet than from many other pesticides.

When exposed in the laboratory to chlorpyrifos at levels roughly estimated from measurements in hives, bee larvae experienced 60% mortality over 6 days, compared with 15% mortality in controls. Adult bees exposed to sub-lethal effects of chlorpyrifos (0.46 ng/bee) exhibited altered behaviors: less walking; more grooming, particularly of the head; more difficulty righting themselves; and unusual abdominal spasms. Chlorpyrifos oxon appears to particularly inhibit acetylcholinesterase in bee gut tissue as opposed to head tissue. Other organophosphate pesticides impaired bee learning and memory of smells in the laboratory.

Regulation

International Law

Chlorpyrifos is not regulated under international law or treaty. Organizations such as PANNA and the NRDC state that chlorpyrifos meets the four criteria (persistence, bioaccumulation, long-range transport, and toxicity) in Annex D of the Stockholm Convention on Persistent Organic Pollutants and should be restricted.

National Regulations

Chlorpyrifos was used to control insect infestations of homes and commercial buildings in Europe until it was banned from sale in 2008.

is restricted from termite control in Singapore as of 2009.

It was banned from residential use in South Africa as of 2010.

In 2010, India barred Dow from commercial activity for 5 years after India's Central Bureau of Investigation found Dow guilty of bribing Indian officials in 2007 to allow the sale of chlorpyrifos.

United States

In the United States, several laws directly or indirectly regulate the use of pesticides. These laws, which are implemented by the EPA, NIOSH, USDA and FDA, include: the Clean Water Act (CWA); the Endangered Species Act (ESA); the Federal Insecticide, Fungicide, and Rodenticide Act (FIFRA); the Federal Food, Drug, and Cosmetic Act (FFDCA); the Comprehensive Environmental Response, Compensation, and Liability Act (CERCLA); and the Emergency Planning and Community Right-to-Know Act (EPCRA). As a pesticide, chlorpyrifos is not regulated under the Toxic Substances Control Act (TSCA).

Chlorpyrifos is sold in restricted-use products for certified pesticide applicators to use in agriculture and other settings, such as golf courses or for mosquito control. It may also be sold in ant and roach baits with childproof packaging. In 2000, manufacturers reached an agreement with the EPA to voluntarily restrict the use of chlorpyrifos in places where children may be exposed, including homes, schools and day care centers.

On August 10, 2015, the Ninth Circuit Court of Appeals ordered the EPA to respond no later than October 2015 to a petition from pesticide activists requesting a chlorpyrifos ban. The court called the EPA's failure to respond for more than eight years (neither granting nor denying the petition) an "egregious" delay. In late October, 2015, the EPA released a proposal to end the use of chlorpyrifos due to a possible risk to certain water supplies. Dow AgroSciences disagreed with the EPA's proposal and said the product had been thoroughly tested for health, safety and environmental effects and said, "no other pesticide has been more thoroughly tested."

Residue

The use of chlorpyrifos in agriculture can leave chemical residue on food commodities. The FFDCA requires EPA to set limits, known as tolerances, for pesticide residue in human food and animal feed products based on risk quotients for acute and chronic exposure from food in humans. These tolerances limit the amount of chlorpyrifos that can be applied to crops. FDA enforces EPA's pesticide tolerances and determines "action levels" for the unintended drift of pesticide residues onto crops without tolerances.

Chlorpyrifos has a tolerance of 0.1 part per million (ppm) residue on all food items unless a different tolerance has been set for that item or chlorpyrifos is not registered for use on that crop. EPA set approximately 112 tolerances pertaining to food products and supplies. In 2006, to reduce childhood exposure, the EPA amended its chlorpyrifos tolerance on apples, grapes and tomatoes, reducing the grape and apple tolerances to 0.01 ppm and eliminating the tolerance on tomatoes. Chlorpyrifos is not allowed on crops such as spinach, squash, carrots, and tomatoes; any chlorpyrifos residue on these crops normally represents chlorpyrifos misuse or spray drift.

Food handling establishments (places where food products are held, processed, prepared or served) are included in the food tolerance of 0.1 ppm for chlorpyrifos. Food handling establishments may use a 0.5% solution of chlorpyrifos solely for spot and/or crack and crevice treatments. Food items are to be removed or protected during treatment. Food handling establishment tolerances may be modified or exempted under FFDCA sec. 408.

Water

Chlorpyrifos in waterways is regulated as a hazardous substance under section 311(b)(2)(A) of the Federal Water Pollution Control Act and falls under the CWA amendments of 1977 and 1978. The regulation is inclusive of all chlorpyrifos isomers and hydrates in any solution or mixture. EPA has not set a drinking water regulatory standard for chlorpyrifos, but has established a drinking water guideline of 2 ug/L.

In 2009, in order to protect threatened salmon and steelhead under CWA and ESA, EPA and National Marine Fisheries Service (NMFS) recommended limits on the use of chlorpyrifos in California, Idaho, Oregon and Washington and requested that manufacturers voluntarily add buffer zones, application limits and fish toxicity to the standard labeling requirements for all chlorpyrifos-based products. Manufacturers rejected the request. In February 2013 in Dow AgroSciences vs NMFS, the Fourth Circuit Court of Appeals vacated EPA's order for these labeling requirements. In August 2014, in the settlement of a suit brought by environmental and fisheries advocacy groups against EPA in the U.S. District Court for the Western District of Washington, EPA agreed to re-instate no-spray stream buffer zones in California, Oregon and Washington, restricting aerial spraying (300 ft.) and ground-based applications (60 ft.) near salmon populations. These buffers will remain until EPA makes a permanent decision in consultation with NMFS.

Reporting

EPCRA designates the chemicals that facilities must report to the Toxics Release Inventory (TRI), based on EPA assessments. Chlorpyrifos is not on the reporting list. It is on the list of hazardous substances under CERCLA (aka the Superfund Act). In the event of an environmental release above its reportable quantity of 1 lb or 0.454 kg, facilities are required to immediately notify the National Response Center (NRC) .

In 1995, Dow paid a $732,000 EPA penalty for not forwarding reports it had received on 249 chlorpyrifos poisoning incidents.

Occupational Exposure

In 1989, OSHA established a workplace permissible exposure limit (PEL) of 0.2 mg/m3 for chlorpyrifos, based on an 8-hour time weighted average (TWA) exposure. However, the rule was remanded by the U.S. Circuit Court of Appeals and no PELs are in place presently.

EPA's Worker Protection Standard requires owners and operators of agricultural businesses to comply with safety protocols for agricultural workers and pesticide handlers (those who mix, load and apply pesticides). For example, in 2005, the EPA filed an administrative complaint against JSH Farms, Inc. (Wapato, Washington) with proposed penalties of $1,680 for using chlorpyrifos in 2004 without proper equipment. An adjacent property was contaminated with chlorpyrifos due to pesticide drift and the property owner suffered from eye and skin irritation.

State Laws

Additional laws and guidelines may apply for individual states. For example, Florida has a drinking water guideline for chlorpyrifos of 21 ug/L. Other states are reviewing chlorpyrifos following the federal government's recommendations for pesticide surveillance.

In 2003, Dow agreed to pay $2 million to New York state, in response to a lawsuit to end Dow's advertising of Dursban as "safe".

Oregon's Department of Environmental Quality added chlorpyrifos to the list of targeted reductions in the Clackamas Subbasin as part of the Columbia River National Strategic Plan, which is based on EPA'S 2006-11 National Strategic Plan.1

In 2008, chlorpyrifos was evaluated for inclusion in California's Proposition 65, a state law that prohibits businesses from discharging substances known to cause birth defects and reproductive harm into the drinking water, but the California's Office of Environmental Health Hazard Assessment decided against the move.

California included regulation limits for chlorpyrifos in waterways and established maximum and continuous concentration limits of 0.025 ppb and 0.015 pbb, respectively.

Pyrethroid

A pyrethroid is an organic compound similar to the natural pyrethrins produced by the flowers of pyrethrums (*Chrysanthemum cinerariaefolium* and *C. coccineum*). Pyrethroids now constitute the majority of commercial household insecticides. In the concentrations used in such products, they may also have insect repellent properties and are generally harmless to human beings in low doses but can harm sensitive individuals. They are usually broken apart by sunlight and the atmosphere in one or two days, and do not significantly affect groundwater quality.

Allethrin

Permethrin

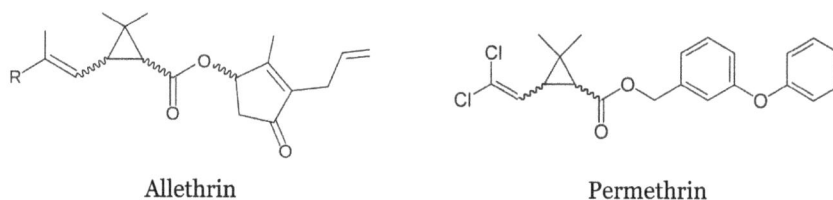

Mode of Action

Pyrethroids are axonic excitoxins, the toxic effects of which are mediated through preventing the closure of the voltage-gated sodium channels in the axonal membranes. The sodium channel is a membrane protein with a hydrophilic interior. This interior is a tiny hole which is shaped precisely to strip away the partially charged water molecules from a sodium ion and create a favorable way for sodium ions to pass through the membrane, enter the axon, and propagate an action potential. When the toxin keeps the channels in their open state, the nerves cannot repolarize, leaving the axonal membrane permanently depolarized, thereby paralyzing the organism.

Pesticide Formulation

Pyrethroids can be combined with the synergist piperonyl butoxide, a known inhibitor of key microsomal cytochrome P450 enzymes from metabolizing the pyrethroid, which increases its efficacy (lethality).

History

Pyrethroids were introduced in the late 1900s (date not accurate) by a team of Rothamsted Research scientists following the elucidation of the structures of pyrethrin I and II by Hermann Staudinger and Leopold Ružička in the 1920s. The pyrethroids represented a major advancement in the chemistry that would synthesize the analog of the natural version found in pyrethrum. Its insecticidal activity has relatively low mammalian toxicity and an unusually fast biodegradation. Their development coincided with the identification of problems with DDT use. Their work consisted firstly of identifying the most active components of pyrethrum, extracted from East African chrysanthemum flowers and long known to have insecticidal properties. Pyrethrum rapidly knocks down flying insects but has negligible persistence — which is good for the environment but gives poor efficacy when applied in the field. Pyrethroids are essentially chemically stabilized forms of natural pyrethrum and belong to IRAC MoA group 3 (they interfere with sodium transport in insect nerve cells).

The *1st generation pyrethroids*, developed in the 1960s, include bioallethrin, tetramethrin, resmethrin and bioresmethrin. They are more active than the natural pyrethrum but are unstable in sunlight. Activity of pyrethrum and 1st generation pyrethroids is often enhanced by addition of the synergist piperonyl butoxide (which itself has some insecticidal activity). With the 91/414/EEC review, many 1st generation compounds have not been included on Annex 1, probably because the market is simply not big enough to warrant the costs of re-registration (rather than any special concerns about safety).

By 1974, the Rothamsted team had discovered a *2nd generation* of more persistent compounds notably: permethrin, cypermethrin and deltamethrin. They are substantially more resistant to degradation by light and air, thus making them suitable for use in agriculture, but they have significantly higher mammalian toxicities. Over the subsequent decades these derivatives were followed with other proprietary compounds such as fenvalerate, lambda-cyhalothrin and beta-cyfluthrin. Most patents have now expired, making these compounds cheap and therefore popular (although permethrin and fenvalerate have not been re-registered under the 91/414/EEC process). One of the less desirable characteristics, especially of 2nd generation pyrethroids is that they can be an irritant to the skin and eyes, so special formulations such as capsule suspensions (CS) have been developed.

Classes of Pyrethroids

The earliest pyrethoids are related to pyrethrin I and II by changing the alcohol group of the ester of chrysanthemic acid. This relatively modest change can lead to substantially altered activities. For example, the 5-benzyl-3-furanyl ester called resmethrin is only weakly toxic to mammals (LD50 (rat, oral) = 2,000 mg/kg) but is 20-50x more effective than natural pyrethrum and is also readily biodegraded. Other commercially important esters include tetramethrin, allethrin, phenothrin, barthrin, dimethrin, and bioresmethrin. Another family of pyrethroids have altered acid fragment together with altered alcohol components. These require more elaborate organic synthesis. Members of this extensive class include the dichlorovinyl and dibromovinyl derivatives. Still others are tefluthrin, fenpropathrin, and bioethanomethrin.

(1R,3R)- or (+)-trans-chrysanthemic acid   Flucythrinate

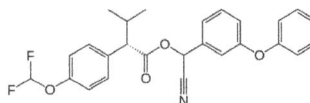

Types

- Allethrin, the first pyrethroid synthesized

- Bifenthrin, active ingredient of *Talstar, Capture, Ortho Home Defense Max*, and *Bifenthrine*

- Cyfluthrin, an active ingredient in Baygon, dichlorovinyl derivative of pyrethrin

- Cypermethrin, including the resolved isomer alpha-cypermethrin, dichlorovinyl derivative of pyrethrin

- Cyphenothrin, active ingredient of K2000 Insect spray sold in Israel and the Palestinian territories

- Deltamethrin, dibromovinyl derivative of pyrethrin

- Esfenvalerate

- Etofenprox

- Fenpropathrin

- Fenvalerate

- Flucythrinate

- Flumethrin

- Imiprothrin

- lambda-Cyhalothrin

- Metofluthrin

- Permethrin, dichlorovinyl derivative of pyrethrin

- Prallethrin, active ingredient in Baygon and All Out (India)

- Resmethrin, active ingredient of *Scourge*

- Silafluofen

- Sumithrin, active ingredient of Anvil

- tau-Fluvalinate

- Tefluthrin

- Tetramethrin

- Tralomethrin

- Transfluthrin, an active ingredient in Baygon

Environmental Effects

Aside from the fact that they are also toxic to beneficial insects such as bees and dragonflies, pyrethroids are toxic to fish and other aquatic organisms. At extremely small levels, such as 4 parts per trillion, pyrethroids are lethal to mayflies, gadflies, and invertebrates that constitute the base of many aquatic and terrestrial food webs.

Pyrethroids have been found to be unaffected by secondary treatment systems at municipal wastewater treatment facilities in California. They appear in the effluent, usually at levels lethal to invertebrates.

Safety and Effectiveness

Earlier studies suggested that most vertebrates have sufficient enzymes for rapid breakdown of pyrethroids, except for cats. Pyrethroids are highly toxic to cats because they do not have glucuronidase, which participates in hepatic detoxifying metabolism pathways.

A recent study, however, suggests that developing mice exposed to deltamethrin (a pyrethroid pesticide) show neurological and behavioral changes resembling Attention-Deficit/Hyperactivity Disorder (ADHD) in humans. In terms of LD_{50} for rats, Tefluthrin is the most toxic at 29 mg/kg.

Anaphylaxis has been reported after pyrethrum exposure, but allergic reaction to pyrethroids has not been documented. Increased sensitivity occurs following repeated exposure to cyanide, which is found in pyrethroids like beta-cyfluthrin (Multiple Chemical Sensitivity Awareness, J. Edward Hill, MD, President & Executive Committee Member, AMA).

Resistance

Up until the 1950s, bedbugs were almost eradicated in the US through the use of DDT. After the use of DDT for this purpose was banned, pyrethroids became more commonly used against bedbugs, but resistant populations have now developed.

Permethrin

Permethrin is a medication and chemical widely used as an insecticide, acaricide, and insect repellent. Permethrin is a first-line treatment for scabies. It is used as a cream.

It belongs to the family of synthetic chemicals called pyrethroids and functions as a neurotoxin, affecting neuron membranes by prolonging sodium channel activation. It is not known to rapidly harm most mammals or birds, but is toxic to fish and cats. In cats it may induce hyperexcitability, tremors, seizures, and death. In general, it has a low mammalian toxicity and is poorly absorbed by skin.

It is on the World Health Organization's List of Essential Medicines, the most important medications needed in a basic health system.

Uses

Permethrin is used:

- as an insecticide

    o in agriculture, to protect crops (lethal for bees)

    o in agriculture, to kill livestock parasites

    o for industrial/domestic insect control

    o in the textile industry to prevent insect attack of woollen products

    o in aviation, the WHO, IHR and ICAO require arriving aircraft be disinsected prior to departure, descent or deplaning in certain countries

    o to treat head lice in humans

- as an insect repellent or insect screen

    o in timber treatment

    o as a personal protective measure (cloth impregnant, used primarily for US military uniforms and mosquito nets)

           ○ in pet flea preventative collars or treatment

  • often in combination with piperonyl butoxide to enhance its effectiveness.

Medical Use

Permethrin is available for topical use as a cream or lotion. It is indicated for the treatment and prevention in exposed individuals of head lice and treatment of scabies.

For treatment of scabies: Adults and children older than 2 months are instructed to apply the cream to the entire body from head to the soles of the feet. Wash off the cream after 8–14 hours. In general, one treatment is curative.

For treatment of head lice: Apply to hair, scalp, and neck after shampooing. Leave in for 10 minutes and rinse. Avoid contact with eyes.

Pest Control

In agriculture, permethrin is mainly used on cotton, wheat, maize, and alfalfa crops. Its use is controversial because, as a broad-spectrum chemical, it kills indiscriminately; as well as the intended pests, it can harm beneficial insects including honey bees, and aquatic life.

Permethrin kills ticks on contact with treated clothing. A method of reducing deer tick populations by treating rodent vectors involves stuffing biodegradable cardboard tubes with permethrin-treated cotton. Mice collect the cotton for lining their nests. Permethrin on the cotton instantly kills any immature ticks feeding on the mice.

Permethrin is used in tropical areas to prevent mosquito-borne disease such as dengue fever and malaria. Mosquito nets used to cover beds may be treated with a solution of permethrin. This increases the effectiveness of the bed net by killing parasitic insects before they are able to find gaps or holes in the net. Military personnel training in malaria-endemic areas may be instructed to treat their uniforms with permethrin, as well.

Permethrin is the most commonly used insecticide worldwide for the protection of wool from keratinophagous insects such as *Tineola bisselliella*.

Side Effects

Permethrin application can cause mild skin irritation and burning. Discontinue use if hypersensitivity occurs.

Safety

Permethrin has little systemic absorption, and is considered safe for topical use in adults and children over the age of 2 months. The FDA has assigned it as pregnancy category B. Animal studies have shown no effects on fertility or teratogenicity, but studies in humans have not been performed. The excretion of permethrin in breastmilk is unknown, and breastfeeding is recommended to be temporarily discontinued during treatment.

Pharmacokinetics

Absorption

Absorption of topical permethrin is minimal. One *in vivo* study demonstrated 0.5% absorption in the first 48 hours based upon excretion of urinary metabolites.

Distribution

Distribution of permethrin has been studied in rat models, with highest amounts accumulating in fat and the brain. This can be explained by the lipophilic nature of the permethrin molecule.

Metabolism

Metabolism of permethrin occurs mainly in the liver, where the molecule undergoes oxidation by the cytochrome P450 system, as well as hydrolysis, into metabolites. Elimination of these metabolites occurs via urinary excretion.

Military use

To better protect soldiers from the risk and annoyance of biting insects, the US and British armies are treating all new uniforms with permethrin.

Stereochemistry

Permethrin has four stereoisomers (two enantiomeric pairs), arising from the two stereocenters in the cyclopropane ring. The *trans* enantiomeric pair is known as transpermethrin.

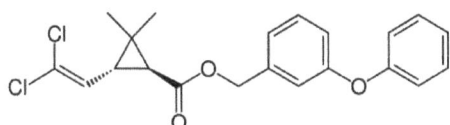

(1S)-trans-acid moiety   one *cis* enantiomer

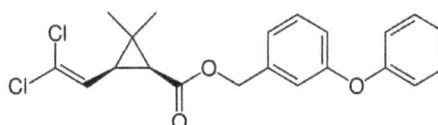

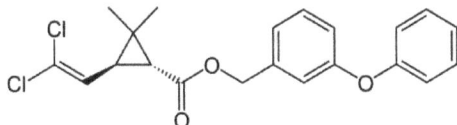

(3S)-trans-acid moiety   the other *cis* enantiomer

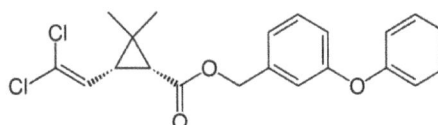

Toxicology and Safety

Permethrin acts as a neurotoxin, slowing down the nervous system through binding to sodium channels. This action is negatively correlated to temperature, thus, in general, showing more acute effects on cold-blooded animals (insects, fish, frogs, etc.) over warm-blooded animals (mammals and birds):

- Permethrin is extremely toxic to fish and aquatic life in general, so extreme care must be taken when using products containing permethrin near water sources.

- Permethrin is also highly toxic to cats, and flea and tick-repellent formulas intended and labeled for (the more resistant) dogs may contain permethrin and cause feline permethrin toxicosis in cats.

- Very high doses have tangible neurotoxic effects on mammals and birds, including human beings.

Permethrin is listed as a "restricted use" substance by the US Environmental Protection Agency (EPA) due to its high toxicity to aquatic organisms, so permethrin and permethrin-contaminated water should be properly disposed. Permethrin is quite stable, having a half life of 51–71 days in an aqueous environment exposed to light. It is also highly persistent in soil.

Human Exposure

According to the Connecticut Department of Public Health, permethrin "has low mammalian toxicity, is poorly absorbed through the skin, and is rapidly inactivated by the body. Skin reactions have been uncommon."

Excessive exposure to permethrin can cause nausea, headache, muscle weakness, excessive salivation, shortness of breath, and seizures. Worker exposure to the chemical can be monitored by measurement of the urinary metabolites, while severe overdose may be confirmed by measurement of permethrin in serum or blood plasma.

Permethrin does not present any notable genotoxicity or immunotoxicity in humans and farm animals, but is classified by the EPA as a likely human carcinogen, based on reproducible studies in which mice fed permethrin developed liver and lung tumors. Carcinogenic action in nasal mucosal cells due to inhalation exposure is suspected, due to observed genotoxicity in human tissue samples, and in rat livers the evidence of increased preneoplastic lesions raises concern over oral exposure.

Animal studies by Bloomquist *et al.* suggest a possible link of permethrin exposure to Parkinson's disease, including very small exposures:

> *2002 study* – "Our studies have documented low-dose effects of permethrin, doses below one-one thousandth of a lethal dose for a mouse, with effects on those brain pathways involved in Parkinson's disease [...] We have found effects consistent with a pre-parkinsonsian condition, but not yet full-blown parkinsonism."

However, a 2007 study by the same researcher concluded "little hazard to humans" existed.

> *2007 study* – "long-term, low-dose exposure to permethrin alone did not cause signs of neurotoxicity to striatal dopaminergic neural terminals, or enhance the effects of MPTP. We conclude that, under typical use conditions, permethrin poses little parkinsonian hazard to humans, including when impregnated into clothing for control of biting flies"

A 2006 study in South Africa found residues of permethrin in breast milk in an area that experienced the use of pyrethroids in small-scale agriculture.

Domestic Animals

Pesticide-grade permethrin is toxic to cats. Many cats die after being given flea treatments intended for dogs, or by contact with dogs having recently been treated with permethrin.

Toxic exposure of permethrin can cause several symptoms, including convulsion, hyperaesthesia, hyperthermia, hypersalivation, and loss of balance and coordination. Exposure to pyrethroid-derived drugs such as permethrin requires treatment by a veterinarian, otherwise the poisoning is often fatal. This intolerance is due to a defect in glucuronosyltransferase, a common detoxification enzyme in other mammals (that also makes the cat intolerant to paracetamol and many essential oils). The use of any external parasiticides based on permethrin is contraindicated for cats. (Cat ecotoxicology : cutaneous 100 mg/kg - oral 200 mg/kg.)

Synthesis

Permethrin was first synthesized in 1973.

Numerous synthetic routes exist for the production of the DV-acid ester precursor. The pathway known as the Kuraray Process uses four steps. In general, the final step in the total synthesis of any of the synthetic pyrethroids is a coupling of a DV-acid ester and an alcohol. In the case of permethrin synthesis, the DV-acid cyclopropanecarboxylic acid, 3-(2,2-dichloroethenyl)-2,2-dimethyl-, ethyl ester, is coupled with the alcohol, m-phenoxybenzyl alcohol, through a transesterification reaction with base. Tetraisopropyl titanate or sodium ethylate may be used as the base.

The alcohol precursor may be prepared in three steps. First, m-cresol, chlorobenzene, sodium hydroxide, potassium hydroxide, and copper chloride react to yield m-phenoxytoluene. Second, oxidation of m-phenoxytoluene over selenium dioxide provides m-phenoxybenzaldehyde. Third, a Cannizzaro reaction of the benzaldehyde in formaldehyde and potassium hydroxide affords the m-phenoxybenzyl alcohol.

Brand Names

It is marketed by Johnson & Johnson under the name Lyclear. In Nordic countries and North America, it is marketed under trade name Nix, often available over the counter.

Insect Repellent

An insect repellent (also commonly called "bug spray") is a substance applied to skin, clothing, or other surfaces which discourages insects (and arthropods in general) from landing or climbing on that surface. Insect repellents help prevent and control the outbreak of insect-borne (and other arthropod-bourne) diseases such as malaria, Lyme disease, dengue fever, bubonic plague, and West Nile fever. Pest animals commonly serving as vectors for disease include insects such as flea, fly, and mosquito; and the arachnid tick.

Some insect repellents are insecticides (bug killers), but most simply discourage insects and send them flying or crawling away. Almost any might kill at a massive dose without reprieve, but classification as an insecticide implies death even at lower doses.

A mosquito coil

Common Insect Repellents

Oil Jar in cow horn for mosquito-repelling pitch oil, a by-product of the distillation of wood tar. Carried in a leather strap on a belt. Råneå, Norrbotten, since 1921 in Nordiska museet, Stockholm.

- Birch tree bark is traditionally made into tar. Combined with another oil (e.g., fish oil) at 1/2 dilution, it is then applied to the skin for repelling mosquitos

- DEET (*N,N*-diethyl-*m*-toluamide)

- Essential oil of the lemon eucalyptus (*Corymbia citriodora*) and its active compound p-menthane-3,8-diol (PMD)

- Icaridin, also known as picaridin, Bayrepel, and KBR 3023

- Nepetalactone, also known as "catnip oil"

- Citronella oil

- Neem oil

- Bog Myrtle (Myrica Gale)

- Dimethyl carbate

- Tricyclodecenyl allyl ether, a compound often found in synthetic perfumes.

- IR3535 (3-[N-Butyl-N-acetyl]-aminopropionic acid, ethyl ester)

- Ethylhexanediol, also known as Rutgers 612 or "6-12 repellent," discontinued in the US in 1991 due to evidence of causing developmental defects in animals

- Dimethyl phthalate, not as common as it once was but still occasionally an active ingredient in commercial insect repellents

- Metofluthrin

- Indalone. Widely used in a "6-2-2" mixture (60% Dimethyl phthalate, 20% Indalone, 20% Ethylhexanediol) during the 1940s and 1950s before the commercial introduction of DEET.

- Permethrin is different in that it is actually a contact insecticide.

- A more recent repellent being currently researched is SS220, which has been shown to provide significantly better protection than DEET.

- Another new and promising group of repellents are the anthranilate-based insect repellents.

Repellent Effectiveness

Synthetic repellents tend to be more effective and/or longer lasting than "natural" repellents. In comparative studies, IR3535 was as effective or better than DEET in protection against mosquitoes. Other sources (official publications of the associations of German physicians as well as of German druggists suggest the contrary and state DEET is still the most efficient substance available and the substance of choice for stays in malaria regions, while IR3535 has little effect. However, some plant-based repellents may provide effective relief as well. Essential oil repellents can be short-lived in their effectiveness, since essential oils can evaporate completely.

A popular post-WWII Australian brand of insect repellent.

A test of various insect repellents by an independent consumer organization found that repellents containing DEET or picaridin are more effective than repellents with "natural" active ingredients. All the synthetics gave almost 100% repellency for the first 2 hours, where the natural repellent products were most effective for the first 30 to 60 minutes, and required reapplication to be effective over several hours.

For protection against mosquitos, the U.S. Centers for Disease Control (CDC) issued a statement in May 2008 recommending equally DEET, picaridin, oil of lemon eucalyptus and IR3535 for skin. Permethrin is recommended for clothing, gear, or bed nets. In an earlier report, the CDC found oil of lemon eucalyptus to be more effective than other plant-based treatments, with a similar effectiveness to low concentrations of DEET. However, a 2006 published study found in both cage and field studies that a product containing 40% oil of lemon eucalyptus was just as effective as products containing high concentrations of DEET. Research has also found that neem oil is mosquito repellent for up to 12 hours. Citronella oil's mosquito repellency has also been verified by research, including effectiveness in repelling *Aedes aegypti*, but requires reapplication after 30 to 60 minutes.

More recently, in 2015, Researchers at New Mexico State University tested 10 commercially available products for their effectiveness at repelling mosquitoes. On the mosquito *Aedes aegypti*, the vector of Zika virus, only one repellent that did not contain DEET had a strong effect for the duration of the 240 minutes test: a lemon eucalyptus oil repellent. All DEET-containing mosquito repellents were active.

There are also products available based on sound production, particularly ultrasound (inaudibly high frequency sounds) which purport to be insect repellents. However, these electronic devices have been shown to be ineffective based on studies done by the EPA and many universities.

Repellent Safety

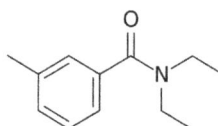

DEET

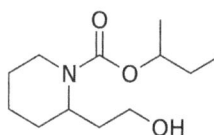

Icaridin

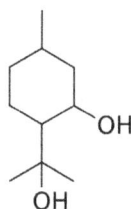

p-Menthane-3,8-diol (PMD)

Regarding safety with insect repellent use on children and pregnant women:

- Children may be at greater risk for adverse reactions to repellents, in part, because their exposure may be greater.

- Keep repellents out of the reach of children.

- Do not allow children to apply repellents to themselves.

- Use only small amounts of repellent on children.

- Do not apply repellents to the hands of young children because this may result in accidental eye contact or ingestion.

- Try to reduce the use of repellents by dressing children in long sleeves and long trousers tucked into boots or socks whenever possible. Use netting over strollers, playpens, etc.

- As with chemical exposures in general, pregnant women should take care to avoid exposures to repellents when practical, as the fetus may be vulnerable.

Some experts also recommend against applying chemicals such as DEET and sunscreen simultaneously since that would increase DEET penetration. Canadian researcher, Xiaochen Gu, a professor at the University of Manitoba's faculty of Pharmacy who led a study about mosquitos, advises that DEET should be applied 30 or more minutes later. Gu also recommends insect repellent sprays instead of lotions which are rubbed into the skin "forcing molecules into the skin".

Regardless of which repellent product used, it is recommended to read the label before use and carefully follow directions. Usage instructions for repellents vary from country to country. Some insect repellents are not recommended for use on younger children.

In the DEET Reregistration Eligibility Decision (RED) the United States Environmental Protection Agency (EPA) reported 14 to 46 cases of potential DEET associated seizures, including 4 deaths. The EPA states: "... it does appear that some cases are likely related to DEET toxicity," but observed that with 30% of the US population using DEET, the likely seizure rate is only about one per 100 million users.

The Pesticide Information Project of Cooperative Extension Offices of Cornell University states that, "Everglades National Park employees having extensive DEET exposure were more likely to have insomnia, mood disturbances and impaired cognitive function than were lesser exposed co-workers".

The EPA states that citronella oil shows little or no toxicity and has been used as a topical insect repellent for 60 years. However, the EPA also states that citronella may irritate skin and cause dermatitis in certain individuals. Canadian regulatory authorities concern with citronella based repellents is primarily based on data-gaps in toxicology, not on incidents.

Within countries of the European Union, implementation of Regulation 98/8/EC, commonly referred to as the Biocidal Products Directive, has severely limited the number and type of insect repellents available to European consumers. Only a small number of active ingredients have been supported by manufacturers in submitting dossiers to the EU Authorities.

In general, only formulations containing DEET, icaridin (sold under the trade name Saltidin and formerly known as Bayrepel or KBR3023), IR3535 (3-[N-Butyl-N-acetyl]-aminopropionic acid, ethyl ester) and Citriodiol (p-menthane-3,8-diol) are available. Most "natural" insect repellents such as citronella, neem oil, and herbal extracts are no longer permitted for sale as insect repellents in the EU; this does not preclude them from being sold for other purposes, as long as the label does not indicate they are a biocide (insect repellent).

Insect Repellents from Natural Sources

Mosquito repellent made from plants.

There are many preparations from naturally occurring sources that have been used as a repellent to certain insects. Some of these act as insecticides while others are only repellent.

- *Achillea alpina* (mosquitos)

- alpha-terpinene (mosquitos)

- Basil

  - Sweet Basil (*Ocimum basilicum*)

- *Callicarpa americana* (Beautyberry)

- Breadfruit (Insect repellent, including mosquitoes)

- Camphor (moths)

- Carvacrol (mosquitos)

- Castor oil (*Ricinus communis*) (mosquitos)

- Catnip oil (*Nepeta* species) (nepetalactone against mosquitos) (WARNING: may attract cats)

- Cedar oil (mosquitos, moths)

- Celery extract (*Apium graveolens*) (mosquitos) In clinical testing an extract of celery was demonstrated to be at least equally effective to 25% DEET, although the commercial availability of such an extract is not known.

- Cinnamon (leaf oil kills mosquito larvae)

- Citronella oil (repels mosquitos)

- Oil of cloves (mosquitos)

- Eucalyptus oil (70%+ eucalyptol), (cineol is a synonym), mosquitos, flies, dust mites)

- Fennel oil (*Foeniculum vulgare*) (mosquitos)

- Garlic (*Allium sativum*) (Mosquito, rice weevil, wheat flour beetle)

- Geranium oil (also known as *Pelargonium graveolens*)

- Lavender (ineffective alone, but measurable effect in certain repellent mixtures)

- Lemon eucalyptus (*Corymbia citriodora*) essential oil and its active ingredient p-menthane-3,8-diol (PMD)

- Lemongrass oil (*Cymbopogon* species) (mosquitos)

  - East-Indian Lemon Grass (*Cymbopogon flexuosus*)

- Marigolds (*Tagetes* species)

- Marjoram (Spider mites *Tetranychus urticae* and *Eutetranychus orientalis*)

- Neem oil (*Azadirachta indica*) (Repels or kills mosquitos, their larvae and a plethora of other insects including those in agriculture)

- Oleic acid, repels bees and ants by simulating the "Smell of death" produced by their decomposing corpses.

- Pennyroyal (*Mentha pulegium*) (mosquitos, fleas), but very toxic to pets.

- Peppermint (*Mentha* x *piperita*) (mosquitos)

- Pyrethrum (from *Chrysanthemum* species, particularly *C. cinerariifolium* and *C. coccineum*)

- Rosemary (*Rosmarinus officinalis*) (mosquitos)

- Spanish Flag (*Lantana camara*) (against Tea Mosquito Bug, *Helopeltis theivora*)

- Tea tree oil from the leaves of *Melaleuca alternifolia*

- Thyme (*Thymus* species) (mosquitos)

- Yellow Nightshade (*Solanum villosum*), berry juice (against *Stegomyia aegypti* mosquitos)

- *Andrographis paniculata* extracts (mosquito)

Inactive Substances – Carriers

In 2002, the *New England Journal of Medicine* published an article that found products containing essential oils such as catnip or geranium oil, when combined with a suitable carrier oil such as soybean, have been found to be effective as natural repellents. This was based on testing done by Johns Hopkins and Cornell Universities. Other commercial products offered for household mosquito "control" include small electrical mats, mosquito repellent vapor, DEET-impregnated wrist bands, mosquito fogging, and mosquito coils containing a form of the chemical allethrin. Mosquito-repellent candles containing citronella oil are sold widely in the U.S. These have been used with mixed reports of success and failure.

Less Effective Methods

Some old studies suggested that the ingestion of large doses of thiamine could be effective as an oral insect repellent against mosquito bites. However, there is now conclusive evidence that thiamin has no efficacy against mosquito bites. Some claim that plants like wormwood or sagewort, lemon balm, lemon grass, lemon thyme and the mosquito plant (Pelargonium) will act against mosquitoes. However, scientists have determined that these plants are "effective" for a limited time only when the leaves are crushed and applied directly to the skin.

There are several, widespread, unproven theories about mosquito control, such as the assertion that vitamin B, in particular B_1 (thiamine), garlic, ultrasonic devices or incense can be used to repel or control mosquitoes. Moreover, manufacturers of "mosquito repelling" ultrasonic devices have been found to be fraudulent, and their devices were deemed "useless" in tests by the UK Consumer magazine *Which?*, and according to a review of scientific studies.

References

- DDT and Its Derivatives: Environmental Aspects, Environmental Health Criteria monograph No. 83, Geneva: World Health Organization, ISBN 92-4-154283-7

- DAVID D (July 4, 2008). "McIntosh residents file suit against Ciba". Archived from the original on November 18, 2010. Retrieved July 7, 2008.

- Michaels D (2008). Doubt is Their Product: How Industry's Assault on Science Threatens Your Health. New York: Oxford University Press. ISBN 978-0-19-530067-3.

- "MFI second page". Malaria Foundation International. Archived from the original on November 18, 2010. Retrieved March 15, 2006.

- "Concern over excessive DDT use in Jiribam fields". The Imphal Free Press. May 5, 2008. Archived from the original on December 6, 2008. Retrieved May 5, 2008.

- "Is DDT still effective and needed in malaria control?". Malaria Foundation International. Archived from the original on November 18, 2010. Retrieved March 15, 2006.

- Moir, John, "New Hurdle for California Condors May Be DDT From Years Ago", The New York Times, November 15, 2010. Retrieved November 15, 2010.

- Walker C, Sibly RM, Hopkin S, Peakall DB (22 December 2005). Principles of ecotoxicology (3rd ed.). Boca Raton, FL: CRC/Taylor & Francis. pp. 300–. ISBN 978-0-8493-3635-5.

- Guillette LJ (2006). "Endocrine Disrupting Contaminants" (PDF). Archived from the original on November 18, 2010. Retrieved February 2, 2007.

- Garrett, Laurie (31 October 1994). The Coming Plague: Newly Emerging Diseases in a World Out of Balance. Farrar, Straus and Giroux. p. 51. ISBN 978-1-4299-5327-6.

- Ghana News Agency (November 17, 2009). "Ministry moves to check unorthodox fishing methods". Ghana News Agency. Archived from the original on November 18, 2010. Retrieved November 18, 2009.

- Finkel, Michael (July 2007). "Malaria". National Geographic. Sarvana A (May 28, 2009). "Bate and Switch: How a free-market magician manipulated two decades of environmental science". Natural Resources New Service. Retrieved June 2, 2009.

- "USAID Health: Infectious Diseases, Malaria, Technical Areas, Prevention and Control, Indoor Residual Spraying". USAID. Archived from the original on November 18, 2010. Retrieved October 14, 2008.

- Hill, Kent R. (2005). "USAID isn't against using DDT in worldwide malaria battle". Archived from the original on March 31, 2006. Retrieved April 3, 2006.

- "Organophosphates FAQs". Centers for Disease Control and Prevention. DHHS Department of Health and Human Services. Retrieved 6 February 2016.

- Yamamoto, Izuru (1999). "Nicotine to Nicotinoids: 1962 to 1997". In Yamamoto, Izuru; Casida, John. Nicotinoid Insecticides and the Nicotinic Acetylcholine Receptor. Tokyo: Springer-Verlag. pp. 3–27. ISBN 443170213X.

- Gervais, J.A.; Luukinen, B.; Buhl, K.; Stone, D. (April 2010). "Imidacloprid Technical Fact Sheet" (PDF). National Pesticide Information Center. Retrieved 12 April 2012.

3

Classification of Pesticides

Pesticides can be classified based on the organisms they act on and the toxicity levels they exhibit. Rated on toxicity, they are categorized into toxicity class I, II, III and IV. Based on the agent they are effective against, they can be grouped into fungicide, herbicide, molluscicide and rodenticide. The chapter explores all the categories of pesticides and also acquaints the reader with toxicity labels.

Toxicity Class

Toxicity class refers to a classification system for pesticides that has been created by a national or international government-related or -sponsored organization. It addresses the acute toxicity of agents such as soil fumigants, fungicides, herbicides, insecticides, miticides, molluscicides, nematicides, or rodenticides.

General Considerations

Assignment to a toxicity class is based typically on results of acute toxicity studies such as the determination of LD_{50} values in animal experiments, notably rodents, via oral, inhaled, or external application. The experimental design measures the acute death rate of an agent. The toxicity class generally does not address issues of other potential harm of the agent, such as bioaccumulation, issues of carcinogenicity, teratogenicity, mutagenic effects, or the impact on reproduction.

Regulating agencies may require that packaging of the agent be labeled with a signal word, a specific warning label to indicate the level of toxicity.

Toxicity Class By Jurisdiction

World Health Organization

The World Health Organization (WHO) names four toxicity classes:

- Class I – a: extremely hazardous

- Class I – b: highly hazardous

- Class II: moderately hazardous

- Class III: slightly hazardous

The system is based on LD50 determination in rats, thus an oral solid agent with an LD50 at 5 mg or less/kg bodyweight is Class I-a, at 5–50 mg/kg Class I-b, at 50–2000 mg/kg Class II, and at more than 2000 mg/kg Class III. Values may differ for liquid oral agents and dermal agents.

European Union

There are eight toxicity classes in the European Union's classification system, which is regulated by Directive 67/548/EEC:

- Class I: very toxic

- Class II: toxic

- Class III: harmful

- Class IV : corrosive

- Class V : irritant

- Class VI : sensitizing

- Class VII : carcinogenic

- Class VIII : mutagenic

Very toxic and toxic substances are marked by the European toxicity symbol.

India

The Indian standardized system of toxicity labels for pesticides uses a 4-color system (red, yellow, blue, green) to plainly label containers with the toxicity class of the contents.

United States

The United States Environmental Protection Agency (EPA) uses four toxicity classes. Classes I to III are required to carry a signal word on the label. Pesticides are regulated in the United States primarily by the Federal Insecticide, Fungicide, and Rodenticide Act (FIFRA).

Toxicity Class I

- most toxic;

- requires signal word: "Danger-Poison", with skull and crossbones symbol, possibly followed by:

  "Fatal if swallowed", "Poisonous if inhaled", "Extremely hazardous by skin contact--rapidly absorbed through skin", or "Corrosive--causes eye damage and severe skin burns"

Class I materials are estimated to be fatal to an adult human at a dose of less than 5 grams (less than a teaspoon).

Toxicity Class II

- moderately toxic

- Signal word: "Warning", possibly followed by:

"Harmful or fatal if swallowed", "Harmful or fatal if absorbed through the skin", "Harmful or fatal if inhaled", or "Causes skin and eye irritation"

Class II materials are estimated to be fatal to an adult human at a dose of 5 to 30 grams.

Toxicity Class III

- slightly toxic

- Signal word: Caution, possibly followed by:

  "Harmful if swallowed", "May be harmful if absorbed through the skin", "May be harmful if inhaled", or "May irritate eyes, nose, throat, and skin"

Class III materials are estimated to be fatal to an adult human at some dose in excess of 30 grams.

Toxicity Class IV

- practically nontoxic

- no Signal Word required since 2002

General Versus Restricted Use

Furthermore, the EPA classifies pesticides into those anybody can apply (*General Use Pesticides*), and those that must be applied by or under the supervision of a certified individual. Application of *Restricted use pesticides* requires that a record of the application be kept.

Fungicide

Fungicides are biocidal chemical compounds or biological organisms used to kill fungi or fungal spores. A fungistatic inhibits their growth. Fungi can cause serious damage in agriculture, resulting in critical losses of yield, quality, and profit. Fungicides are used both in agriculture and to fight fungal infections in animals. Chemicals used to control oomycetes, which are not fungi, are also referred to as fungicides, as oomycetes use the same mechanisms as fungi to infect plants.

Fungicides can either be contact, translaminar or systemic. Contact fungicides are not taken up into the plant tissue and protect only the plant where the spray is deposited. Translaminar fungicides redistribute the fungicide from the upper, sprayed leaf surface to the lower, unsprayed surface. Systemic fungicides are taken up and redistributed through the xylem vessels. Few fungicides move to all parts of a plant. Some are locally systemic, and some move upwardly.

Most fungicides that can be bought retail are sold in a liquid form. A very common active ingredient is sulfur, present at 0.08% in weaker concentrates, and as high as 0.5% for more potent fungicides. Fungicides in powdered form are usually around 90% sulfur and are very toxic. Other active ingredients in fungicides include neem oil, rosemary oil, jojoba oil, the bacterium *Bacillus subtilis*, and the beneficial fungus *Ulocladium oudemansii*.

Fungicide residues have been found on food for human consumption, mostly from post-harvest treatments. Some fungicides are dangerous to human health, such as vinclozolin, which has now been removed from use. Ziram is also a fungicide that is thought to be toxic to humans if exposed to chronically. A number of fungicides are also used in human health care.

Natural Fungicides

Plants and other organisms have chemical defenses that give them an advantage against microorganisms such as fungi. Some of these compounds can be used as fungicides:

- Tea tree oil
- Cinnamaldehyde
- Citronella oil
- Jojoba oil
- Nimbin
- Oregano oil
- Rosemary oil
- Monocerin
- Milk

Whole live or dead organisms that are efficient at killing or inhibiting fungi can sometimes be used as fungicides:

- *Bacillus subtilis*
- *Ulocladium oudemansii*
- Kelp (powdered dried kelp is fed to cattle to help prevent fungal infection)
- *Ampelomyces quisqualis*

Resistance

Pathogens respond to the use of fungicides by evolving resistance. In the field several mechanisms of resistance have been identified. The evolution of fungicide resistance can be gradual or sudden. In qualitative or discrete resistance, a mutation (normally to a single gene) produces a race of a fungus with a high degree of resistance. Such resistant varieties also tend to show stability, persisting after the fungicide has been removed from the market. For example, sugar beet leaf blotch remains resistant to azoles years after they were no longer used for control of the disease. This is because such mutations often have a high selection pressure when the fungicide is used, but there is low selection pressure to remove them in the absence of the fungicide.

In instances where resistance occurs more gradually, a shift in sensitivity in the pathogen to the fungicide can be seen. Such resistance is polygenic – an accumulation of many mutations in different genes, each having a small additive effect. This type of resistance is known as quantitative or continuous resistance. In this kind of resistance, the pathogen population will revert to a sensitive state if the fungicide is no longer applied.

Little is known about how variations in fungicide treatment affect the selection pressure to evolve resistance to that fungicide. Evidence shows that the doses that provide the most control of the disease also provide the largest selection pressure to acquire resistance, and that lower doses decrease the selection pressure.

In some cases when a pathogen evolves resistance to one fungicide, it automatically obtains resistance to others – a phenomenon known as cross resistance. These additional fungicides are normally of the same chemical family or have the same mode of action, or can be detoxified by the same mechanism. Sometimes negative cross resistance occurs, where resistance to one chemical class of fungicides leads to an increase in sensitivity to a different chemical class of fungicides. This has been seen with carbendazim and diethofencarb.

There are also recorded incidences of the evolution of multiple drug resistance by pathogens – resistance to two chemically different fungicides by separate mutation events. For example, *Botrytis cinerea* is resistant to both azoles and dicarboximide fungicides.

There are several routes by which pathogens can evolve fungicide resistance. The most common mechanism appears to be alteration of the target site, in particular as a defence against single site of action fungicides. For example, Black Sigatoka, an economically important pathogen of banana, is resistant to the QoI fungicides, due to a single nucleotide change resulting in the replacement of one amino acid (glycine) by another (alanine) in the target protein of the QoI fungicides, cytochrome b. It is presumed that this disrupts the binding of the fungicide to the protein, rendering the fungicide ineffective. Upregulation of target genes can also render the fungicide ineffective. This is seen in DMI-resistant strains of *Venturia inaequalis*.

Resistance to fungicides can also be developed by efficient efflux of the fungicide out of the cell. *Septoria tritici* has developed multiple drug resistance using this mechanism. The pathogen had 5 ABC-type transporters with overlapping substrate specificities that together work to pump toxic chemicals out of the cell.

In addition to the mechanisms outlined above, fungi may also develop metabolic pathways that circumvent the target protein, or acquire enzymes that enable metabolism of the fungicide to a harmless substance.

Fungicide Resistance Management

The fungicide resistance action committee (FRAC) has several recommended practices to try to avoid the development of fungicide resistance, especially in at-risk fungicides including *Strobilurins* such as azoxystrobin.

Products should not be used in isolation, but rather as mixture, or alternate sprays, with another fungicide with a different mechanism of action. The likelihood of the pathogen's developing resistance is greatly decreased by the fact that any resistant isolates to one fungicide will be killed by the other; in other words, two mutations would be required rather than just one. The effectiveness of this technique can be demonstrated by Metalaxyl, a phenylamide fungicide. When used as the sole product in Ireland to control potato blight (*Phytophthora infestans*), resistance developed within one growing season. However, in countries like the UK where it was marketed only as a mixture, resistance problems developed more slowly.

Fungicides should be applied only when absolutely necessary, especially if they are in an at-risk group. Lowering the amount of fungicide in the environment lowers the selection pressure for resistance to develop.

Manufacturers' doses should always be followed. These doses are normally designed to give the right balance between controlling the disease and limiting the risk of resistance development. Higher doses increase the selection pressure for single-site mutations that confer resistance, as all strains but those that carry the mutation will be eliminated, and thus the resistant strain will propagate. Lower doses greatly increase the risk of polygenic resistance, as strains that are slightly less sensitive to the fungicide may survive.

It is also recommended that where possible fungicides are used only in a protective manner, rather than to try to cure already-infected crops. Far fewer fungicides have curative/eradicative ability than protectant. Thus, fungicide preparations advertised as having curative action may have only one active chemical; a single fungicide acting in isolation increases the risk of fungicide resistance.

It is better to use an integrative pest management approach to disease control rather than relying on fungicides alone. This involves the use of resistant varieties and hygienic practices, such as the removal of potato discard piles and stubble on which the pathogen can overwinter, greatly reducing the titre of the pathogen and thus the risk of fungicide resistance development.

Herbicide

Weeds controlled with herbicide

Herbicide(s), also commonly known as weedkillers, are chemical substances used to control unwanted plants. Selective herbicides control specific weed species, while leaving the desired crop relatively unharmed, while non-selective herbicides (sometimes called "total weedkillers" in commercial products) can be used to clear waste ground, industrial and construction sites, railways and railway embankments as they kill all plant material with which they come into contact. Apart from selective/non-selective, other important distinctions include *persistence* (also known as *residual action*: how long the product stays in place and remains active), *means of uptake* (whether it is absorbed by above-ground foliage only, through the roots, or by other means), and *mechanism of action* (how it works). Historically, products such as common salt and other metal salts were used as herbicides, however these have gradually fallen out of favor and in some countries a number of these are banned due to their persistence in soil, and toxicity and groundwater contamination concerns. Herbicides have also been used in warfare and conflict.

Modern herbicides are often synthetic mimics of natural plant hormones which interfere with growth of the target plants. The term organic herbicide has come to mean herbicides intended for organic farming; these are often less efficient and more costly than synthetic herbicides and are based on natural materials. Some plants also produce their own natural herbicides, such as the genus *Juglans* (walnuts), or the tree of heaven; such action of natural herbicides, and other related chemical interactions, is called allelopathy. Due to herbicide resistance - a major concern in agriculture - a number of products also combine herbicides with different means of action.

In the US in 2007, about 83% of all herbicide usage, determined by weight applied, was in agriculture. In 2007, world pesticide expenditures totaled about $39.4 billion; herbicides were about 40% of those sales and constituted the biggest portion, followed by insecticides, fungicides, and other types. Smaller quantities are used in forestry, pasture systems, and management of areas set aside as wildlife habitat.

History

Prior to the widespread use of chemical herbicides, cultural controls, such as altering soil pH, salinity, or fertility levels, were used to control weeds. Mechanical control (including tillage) was also (and still is) used to control weeds.

First Herbicides

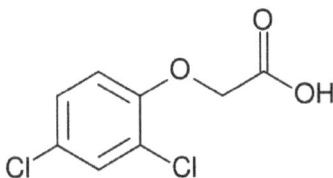

2,4-D, the first chemical herbicide, was discovered during the Second World War.

Although research into chemical herbicides began in the early 20th century, the first major breakthrough was the result of research conducted in both the UK and the US during the Second World War into the potential use of agents as biological weapons. The first modern herbicide, 2,4-D, was first discovered and synthesized by W. G. Templeman at Imperial Chemical Industries. In 1940, he showed that "Growth substances applied appropriately would kill certain broad-leaved weeds in cereals without harming the crops." By 1941, his team succeeded in synthesizing the chemical. In the same year, Pokorny in the US achieved this as well.

Independently, a team under Juda Hirsch Quastel, working at the Rothamsted Experimental Station made the same discovery. Quastel was tasked by the Agricultural Research Council (ARC) to discover methods for improving crop yield. By analyzing soil as a dynamic system, rather than an inert substance, he was able to apply techniques such as perfusion. Quastel was able to quantify the influence of various plant hormones, inhibitors and other chemicals on the activity of microorganisms in the soil and assess their direct impact on plant growth. While the full work of the unit remained secret, certain discoveries were developed for commercial use after the war, including the 2,4-D compound.

When it was commercially released in 1946, it triggered a worldwide revolution in agricultural output and became the first successful selective herbicide. It allowed for greatly enhanced weed control in wheat, maize (corn), rice, and similar cereal grass crops, because it kills dicots (broadleaf plants), but not most monocots (grasses). The low cost of 2,4-D has led to continued usage today, and it remains one of the most commonly used herbicides in the world. Like other acid herbicides, current formulations use either an amine salt (often trimethylamine) or one of many esters of the parent compound. These are easier to handle than the acid.

Further Discoveries

The triazine family of herbicides, which includes atrazine, were introduced in the 1950s; they have the current distinction of being the herbicide family of greatest concern regarding groundwater contamination. Atrazine does not break down readily (within a few weeks) after being applied to soils of above neutral pH. Under alkaline soil conditions, atrazine may be carried into the soil profile as far as the water table by soil water following rainfall causing the aforementioned contamination. Atrazine is thus said to have "carryover", a generally undesirable property for herbicides.

Glyphosate (Roundup) was introduced in 1974 for nonselective weed control. Following the development of glyphosate-resistant crop plants, it is now used very extensively for selective weed control in growing crops. The pairing of the herbicide with the resistant seed contributed to the consolidation of the seed and chemistry industry in the late 1990s.

Many modern chemical herbicides used in agriculture and gardening are specifically formulated to decompose within a short period after application. This is desirable, as it allows crops and plants to be planted afterwards, which could otherwise be affected by the herbicide. However, herbicides with low residual activity (i.e., that decompose quickly) often do not provide season-long weed control and do not ensure that weed roots are killed beneath construction and paving (and cannot emerge destructively in years to come), therefore there remains a role for weedkiller with high levels of persistence in the soil.

Terminology

Herbicides are classified/grouped in various ways e.g. according to the activity, timing of application, method of application, mechanism of action, chemical family. This gives rise to a considerable level of terminology related to herbicides and their use.

Intended Outcome

- Control is the destruction of unwanted weeds, or the damage of them to the point where they are no longer competitive with the crop.

- Suppression is incomplete control still providing some economic benefit, such as reduced competition with the crop.

- Crop safety, for selective herbicides, is the relative absence of damage or stress to the crop. Most selective herbicides cause some visible stress to crop plants.

- Defoliant, similar to herbicides, but designed to remove foliage (leaves) rather than kill the plant.

Selectivity (All Plants or Specific Plants)

- Selective herbicides: They control or suppress certain plants without affecting the growth of other plants species. Selectivity may be due to translocation, differential absorption, physical (morphological) or physiological differences between plant species. 2,4-D, mecoprop, dicamba control many broadleaf weeds but remain ineffective against turfgrasses.

- Non-selective herbicides: These herbicides are not specific in acting against certain plant species and control all plant material with which they come into contact. They are used to clear industrial sites, waste ground, railways and railway embankments. Paraquat, glufosinate, glyphosate are non-selective herbicides.

Timing of Application

- Preplant: Preplant herbicides are nonselective herbicides applied to soil before planting. Some preplant herbicides may be mechanically incorporated into the soil. The objective for incorporation is to prevent dissipation through photodecomposition and/or volatility. The herbicides kill weeds as they grow through the herbicide treated zone. Volatile herbicides have to be incorporated into the soil before planting the pasture. Agricultural crops grown in soil treated with a preplant herbicide include tomatoes, corn, soybeans and strawberries. Soil fumigants like metam-sodium and dazomet are in use as preplant herbicides.

- Preemergence: Preemergence herbicides are applied before the weed seedlings emerge through the soil surface. Herbicides do not prevent weeds from germinating but they kill weeds as they grow through the herbicide treated zone by affecting the cell division in the emerging seedling. Dithopyr and pendimethalin are preemergence herbicides. Weeds that have already emerged before application or activation are not affected by pre-herbicides as their primary growing point escapes the treatment.

- Postemergence: These herbicides are applied after weed seedlings have emerged through the soil surface. They can be foliar or root absorbed, selective or nonselective, contact or systemic. Application of these herbicides is avoided during rain because the problem of being washed off to the soil makes it ineffective. 2,4-D is a selective, systemic, foliar absorbed postemergence herbicide.

Method of Application

- Soil applied: Herbicides applied to the soil are usually taken up by the root or shoot of the emerging seedlings and are used as preplant or preemergence treatment. Several factors influence the effectiveness of soil-applied herbicides. Weeds absorb herbicides by both passive and active mechanism. Herbicide adsorption to soil colloids or organic matter often reduces its amount available for weed absorption. Positioning of herbicide in correct layer of soil is very important, which can be achieved mechanically and by rainfall. Herbicides on the soil surface are subjected to several processes that reduce their availability. Volatility and photolysis are two common processes that reduce the availability of herbicides. Many soil applied herbicides are absorbed through plant shoots while they are still underground leading to their death or injury. EPTC and trifluralin are soil applied herbicides.

- Foliar applied: These are applied to portion of the plant above the ground and are absorbed by exposed tissues. These are generally postemergence herbicides and can either be translocated (systemic) throughout the plant or remain at specific site (contact). External barriers of plants like cuticle, waxes, cell wall etc. affect herbicide absorption and action. Glyphosate, 2,4-D and dicamba are foliar applied herbicide.

Persistence

- Residual activity: A herbicide is described as having low residual activity if it is neutralized within a short time of application (within a few weeks or months) - typically this is due to rainfall, or by reactions in the soil. A herbicide described as having high residual activity will remains potent for a long term in the soil. For some compounds, the residual activity can leave the ground almost permanently barren.

Mechanism of Action

Herbicides are often classified according to their site of action, because as a general rule, herbicides within the same site of action class will produce similar symptoms on susceptible plants. Classification based on site of action of herbicide is comparatively better as herbicide resistance management can be handled more properly and effectively. Classification by mechanism of action (MOA) indicates the first enzyme, protein, or biochemical step affected in the plant following application.

List of Mechanisms Found in Modern Herbicides

- ACCase inhibitors compounds kill grasses. Acetyl coenzyme A carboxylase (ACCase) is part of the first step of lipid synthesis. Thus, ACCase inhibitors affect cell membrane production in the meristems of the grass plant. The ACCases of grasses are sensitive to these herbicides, whereas the ACCases of dicot plants are not.

- ALS inhibitors: the acetolactate synthase (ALS) enzyme (also known as acetohydroxyacid synthase, or AHAS) is the first step in the synthesis of the branched-chain amino acids (valine, leucine, and isoleucine). These herbicides slowly starve affected plants of these amino acids, which eventually leads to inhibition of DNA synthesis. They affect grasses and dicots alike. The ALS inhibitor family includes various sulfonylureas (such as Flazasulfuron and Metsulfuron-methyl), imidazolinones, triazolopyrimidines, pyrimidinyl oxybenzoates, and sulfonylamino carbonyl triazolinones. The ALS biological pathway exists only in plants and not animals, thus making the ALS-inhibitors among the safest herbicides.

- EPSPS inhibitors: The enolpyruvylshikimate 3-phosphate synthase enzyme EPSPS is used in the synthesis of the amino acids tryptophan, phenylalanine and tyrosine. They affect grasses and dicots alike. Glyphosate (Roundup) is a systemic EPSPS inhibitor inactivated by soil contact.

- Synthetic auxins inaugurated the era of organic herbicides. They were discovered in the 1940s after a long study of the plant growth regulator auxin. Synthetic auxins mimic this plant hormone. They have several points of action on the cell membrane, and are effective in the control of dicot plants. 2,4-D is a synthetic auxin herbicide.

- Photosystem II inhibitors reduce electron flow from water to $NADPH2+$ at the photochemical step in photosynthesis. They bind to the Qb site on the D1 protein, and prevent quinone from binding to this site. Therefore, this group of compounds causes electrons to accumulate on chlorophyll molecules. As a consequence, oxidation reactions in excess of those normally tolerated by the cell occur, and the plant dies. The triazine herbicides (including atrazine) and urea derivatives (diuron) are photosystem II inhibitors.

- Photosystem I inhibitors steal electrons from the normal pathway through FeS to Fdx to NADP leading to direct discharge of electrons on oxygen. As a result, reactive oxygen species are produced and oxidation reactions in excess of those normally tolerated by the cell occur, leading to plant death. Bipyridinium herbicides (such as diquat and paraquat) inhibit the Fe-S – Fdx step of that chain, while diphenyl ether herbicides (such as nitrofen, nitrofluorfen, and acifluorfen) inhibit the Fdx – NADP step.

- HPPD inhibitors inhibit 4-Hydroxyphenylpyruvate dioxygenase, which are involved in tyrosine breakdown. Tyrosine breakdown products are used by plants to make carotenoids, which protect chlorophyll in plants from being destroyed by sunlight. If this happens, the plants turn white due to complete loss of chlorophyll, and the plants die. Mesotrione and sulcotrione are herbicides in this class; a drug, nitisinone, was discovered in the course of developing this class of herbicides.

Herbicide Group (Labeling)

One of the most important methods for preventing, delaying, or managing resistance is to reduce the reliance on a single herbicide mode of action. To do this, farmers must know the mode of action for the herbicides they intend to use, but the relatively complex nature of plant biochemistry makes this difficult to determine. Attempts were made to simplify the understanding of herbicide mode of action by developing a classification system that grouped herbicides by mode of action. Eventually the Herbicide Resistance Action Committee (HRAC) and the Weed Science Society of America (WSSA) developed a classification system. The WSSA and HRAC systems differ in the group designation. Groups in the WSSA and the HRAC systems are designated by numbers and letters, respectively. The goal for adding the "Group" classification and mode of action to the herbicide product label is to provide a simple and practical approach to deliver the information to users. This information will make it easier to develop educational material that is consistent and effective. It should increase user's awareness of herbicide mode of action and provide more accurate recommendations for resistance management. Another goal is to make it easier for users to keep records on which herbicide mode of actions are being used on a particular field from year to year.

Chemical Family

Detailed investigations on chemical structure of the active ingredients of the registered herbicides showed that some moieties (moiety is a part of a molecule that may include either whole functional groups or parts of functional groups as substructures; a functional group has similar chemical properties whenever it occurs in different compounds) have the same mechanisms of action. According to Forouzesh *et al.* 2015, these moieties have been assigned to the names of chemical families and active ingredients are then classified within the chemical families accordingly. Knowing about herbicide chemical family grouping could serve as a short-term strategy for managing resistance to site of action.

Use and Application

Most herbicides are applied as water-based sprays using ground equipment. Ground equipment varies in design, but large areas can be sprayed using self-propelled sprayers equipped with long booms, of 60 to 120 feet (18 to 37 m) with spray nozzles spaced every 20–30 inches (510–760 mm) apart. Towed, handheld, and even horse-drawn sprayers are also used. On large areas, herbicides may also at times be applied aerially using helicopters or airplanes, or through irrigation systems (known as chemigation).

Herbicides being sprayed from the spray arms of a tractor in North Dakota.

A further method of herbicide application developed around 2010, involves ridding the soil of its active weed seed bank rather than just killing the weed. This can successfully treat annual plants but not perennials. Researchers at the Agricultural Research Service found that the application of herbicides to fields late in the weeds' growing season greatly reduces their seed production, and therefore fewer weeds will return the following season. Because most weeds are annuals, their seeds will only survive in soil for a year or two, so this method will be able to destroy such weeds after a few years of herbicide application.

Weed-wiping may also be used, where a wick wetted with herbicide is suspended from a boom and dragged or rolled across the tops of the taller weed plants. This allows treatment of taller grassland weeds by direct contact without affecting related but desirable shorter plants in the grassland sward beneath. The method has the benefit of avoiding spray drift. In Wales, a scheme offering free weed-wiper hire was launched in 2015 in an effort to reduce the levels of MCPA in water courses.

Misuse and Misapplication

Herbicide volatilisation or spray drift may result in herbicide affecting neighboring fields or plants, particularly in windy conditions. Sometimes, the wrong field or plants may be sprayed due to error.

Use Politically, Militarily, and In Conflict

Health and Environmental Effects

Herbicides have widely variable toxicity in addition to acute toxicity from occupational exposure levels.

Some herbicides cause a range of health effects ranging from skin rashes to death. The pathway of attack can arise from intentional or unintentional direct consumption, improper application resulting in the herbicide coming into direct contact with people or wildlife, inhalation of aerial sprays, or food consumption prior to the labeled preharvest interval. Under some conditions, certain herbicides can be transported via leaching or surface runoff to contaminate groundwater or distant surface water sources. Generally, the conditions that promote herbicide transport include intense storm events (particularly shortly after application) and soils with limited capacity to adsorb or retain the herbicides. Herbicide properties that increase likelihood of transport include persistence (resistance to degradation) and high water solubility.

Phenoxy herbicides are often contaminated with dioxins such as TCDD; research has suggested such contamination results in a small rise in cancer risk after occupational exposure to these herbicides. Triazine exposure has been implicated in a likely relationship to increased risk of breast cancer, although a causal relationship remains unclear.

Herbicide manufacturers have at times made false or misleading claims about the safety of their products. Chemical manufacturer Monsanto Company agreed to change its advertising after pressure from New York attorney general Dennis Vacco; Vacco complained about misleading claims that its spray-on glyphosate-based herbicides, including Roundup, were safer than table salt and "practically non-toxic" to mammals, birds, and fish (though proof that this was ever said is hard to find). Roundup is toxic and has resulted in death after being ingested in quantities ranging from 85 to 200 ml, although it has also been ingested in quantities as large as 500 ml with only mild or moderate symptoms. The manufacturer of Tordon 101 (Dow AgroSciences, owned by the Dow Chemical Company) has claimed Tordon 101 has no effects on animals and insects, in spite of evidence of strong carcinogenic activity of the active ingredient Picloram in studies on rats.

The risk of Parkinson's disease has been shown to increase with occupational exposure to herbicides and pesticides. The herbicide paraquat is suspected to be one such factor.

All commercially sold, organic and nonorganic herbicides must be extensively tested prior to approval for sale and labeling by the Environmental Protection Agency. However, because of the large number of herbicides in use, concern regarding health effects is significant. In addition to health effects caused by herbicides themselves, commercial herbicide mixtures often contain other chemicals, including inactive ingredients, which have negative impacts on human health.

Ecological Effects

Commercial herbicide use generally has negative impacts on bird populations, although the impacts are highly variable and often require field studies to predict accurately. Laboratory studies have at times overestimated negative impacts on birds due to toxicity, predicting serious problems that were not observed in the field. Most observed effects are due not to toxicity, but to habitat changes and the decreases in abundance of species on which birds rely for food or shelter. Herbicide use in silviculture, used to favor certain types of growth following clearcutting, can cause significant drops in bird populations. Even when herbicides which have low toxicity to birds are used, they decrease the abundance of many types of vegetation on which the birds rely. Herbicide use in agriculture in Britain has been linked to a decline in seed-eating bird species which rely on the weeds killed by the herbicides. Heavy use of herbicides in neotropical agricultural areas has been one of many factors implicated in limiting the usefulness of such agricultural land for wintering migratory birds.

Frog populations may be affected negatively by the use of herbicides as well. While some studies have shown that atrazine may be a teratogen, causing demasculinization in male frogs, the U.S. Environmental Protection Agency (EPA) and its independent Scientific Advisory Panel (SAP) examined all available studies on this topic and concluded that "atrazine does not adversely affect amphibian gonadal development based on a review of laboratory and field studies."

Scientific Uncertainty of Full Extent of Herbicide Effects

The health and environmental effects of many herbicides is unknown, and even the scientific community often disagrees on the risk. For example, a 1995 panel of 13 scientists reviewing studies on the carcinogenicity of 2,4-D had divided opinions on the likelihood 2,4-D causes cancer in humans. As of 1992, studies on phenoxy herbicides were too few to accurately assess the risk of many types of cancer from these herbicides, even though evidence was stronger that exposure to these herbicides is associated with increased risk of soft tissue sarcoma and non-Hodgkin lymphoma. Furthermore, there is some suggestion that herbicides can play a role in sex reversal of certain organisms that experience temperature-dependent sex determination, which could theoretically alter sex ratios.

Resistance

Weed resistance to herbicides has become a major concern in crop production worldwide. Resistance to herbicides is often attributed to lack of rotational programmes of herbicides and to continuous applications of herbicides with the same sites of action. Thus, a true understanding of the sites of action of herbicides is essential for strategic planning of herbicide-based weed control.

Plants have developed resistance to atrazine and to ALS-inhibitors, and more recently, to glyphosate herbicides. Marestail is one weed that has developed glyphosate resistance. Glyphosate-resistant weeds are present in the vast majority of soybean, cotton and corn farms in some U.S. states. Weeds that can resist multiple other herbicides are spreading. Few new herbicides are near commercialization, and none with a molecular mode of action for which there is no resistance. Because most herbicides could not kill all weeds, farmers rotated crops and herbicides to stop resistant weeds. During its initial years, glyphosate was not subject to resistance and allowed farmers to reduce the use of rotation.

A family of weeds that includes waterhemp (Amaranthus rudis) is the largest concern. A 2008-9 survey of 144 populations of waterhemp in 41 Missouri counties revealed glyphosate resistance in 69%. Weeds from some 500 sites throughout Iowa in 2011 and 2012 revealed glyphosate resistance in approximately 64% of waterhemp samples. The use of other killers to target "residual" weeds has become common, and may be sufficient to have stopped the spread of resistance From 2005 through 2010 researchers discovered 13 different weed species that had developed resistance to glyphosate. But since then only two more have been discovered. Weeds resistant to multiple herbicides with completely different biological action modes are on the rise. In Missouri, 43% of samples were resistant to two different herbicides; 6% resisted three; and 0.5% resisted four. In Iowa 89% of waterhemp samples resist two or more herbicides, 25% resist three, and 10% resist five.

For southern cotton, herbicide costs has climbed from between $50 and $75 per hectare a few years ago to about $370 per hectare in 2013. Resistance is contributing to a massive shift away from growing cotton; over the past few years, the area planted with cotton has declined by 70% in Arkansas and by 60% in Tennessee. For soybeans in Illinois, costs have risen from about $25 to $160 per hectare.

Dow, Bayer CropScience, Syngenta and Monsanto are all developing seed varieties resistant to herbicides other than glyphosate, which will make it easier for farmers to use alternative weed killers. Even though weeds have already evolved some resistance to those herbicides, Powles says the new seed-and-herbicide combos should work well if used with proper rotation.

Biochemistry of Resistance

Resistance to herbicides can be based on one of the following biochemical mechanisms:

- Target-site resistance: This is due to a reduced (or even lost) ability of the herbicide to bind to its target protein. The effect usually relates to an enzyme with a crucial function in a metabolic pathway, or to a component of an electron-transport system. Target-site resistance may also be caused by an overexpression of the target enzyme (via gene amplification or changes in a gene promoter).

- Non-target-site resistance: This is caused by mechanisms that reduce the amount of herbicidal active compound reaching the target site. One important mechanism is an enhanced metabolic detoxification of the herbicide in the weed, which leads to insufficient amounts of the active substance reaching the target site. A reduced uptake and translocation, or sequestration of the herbicide, may also result in an insufficient herbicide transport to the target site.

- Cross-resistance: In this case, a single resistance mechanism causes resistance to several herbicides. The term target-site cross-resistance is used when the herbicides bind to the same target site, whereas non-target-site cross-resistance is due to a single non-target-site mechanism (e.g., enhanced metabolic detoxification) that entails resistance across herbicides with different sites of action.

- Multiple resistance: In this situation, two or more resistance mechanisms are present within individual plants, or within a plant population.

Resistance Management

Worldwide experience has been that farmers tend to do little to prevent herbicide resistance developing, and only take action when it is a problem on their own farm or neighbor's. Careful observation is important so that any reduction in herbicide efficacy can be detected. This may indicate evolving resistance. It is vital that resistance is detected at an early stage as if it becomes an acute, whole-farm problem, options are more limited and greater expense is almost inevitable. Table 1 lists factors which enable the risk of resistance to be assessed. An essential pre-requisite for confirmation of resistance is a good diagnostic test. Ideally this should be rapid, accurate, cheap and accessible. Many diagnostic tests have been developed, including glasshouse pot assays, petri dish assays and chlorophyll fluorescence. A key component of such tests is that the response of the suspect population to a herbicide can be compared with that of known susceptible and resistant standards under controlled conditions. Most cases of herbicide resistance are a consequence of the repeated use of herbicides, often in association with crop monoculture and reduced cultivation practices. It is necessary, therefore, to modify these practices in order to prevent or delay the onset of resistance or to control existing resistant populations. A key objective should be the reduction in selection pressure. An integrated weed management (IWM) approach is required, in which as many tactics as possible are used to combat weeds. In this way, less reliance is placed on herbicides and so selection pressure should be reduced.

Optimising herbicide input to the economic threshold level should avoid the unnecessary use of herbicides and reduce selection pressure. Herbicides should be used to their greatest potential by ensuring that the timing, dose, application method, soil and climatic conditions are optimal for good activity. In the UK, partially resistant grass weeds such as *Alopecurus myosuroides* (blackgrass) and *Avena* spp. (wild oat) can often be controlled adequately when herbicides are applied at the 2-3 leaf stage, whereas later applications at the 2-3 tiller stage can fail badly. Patch spraying, or applying herbicide to only the badly infested areas of fields, is another means of reducing total herbicide use.

Table 1. Agronomic factors influencing the risk of herbicide resistance development

Factor	Low risk	High risk
Cropping system	Good rotation	Crop monoculture
Cultivation system	Annual ploughing	Continuous minimum tillage
Weed control	Cultural only	Herbicide only
Herbicide use	Many modes of action	Single modes of action
Control in previous years	Excellent	Poor
Weed infestation	Low	High
Resistance in vicinity	Unknown	Common

Approaches to Treating Resistant Weeds

Alternative Herbicides

When resistance is first suspected or confirmed, the efficacy of alternatives is likely to be the first consideration. The use of alternative herbicides which remain effective on resistant populations can be a successful strategy, at least in the short term. The effectiveness of alternative herbicides will be highly dependent on the extent of cross-resistance. If there is resistance to a single group of herbicides, then the use of herbicides from other groups may provide a simple and effective solution, at least in the short term. For example, many triazine-resistant weeds have been readily controlled by the use of alternative herbicides such as dicamba or glyphosate. If resistance extends to more than one herbicide group, then choices are more limited. It should not be assumed that resistance will automatically extend to all herbicides with the same mode of action, although it is wise to assume this until proved otherwise. In many weeds the degree of cross-resistance between the five groups of ALS inhibitors varies considerably. Much will depend on the resistance mechanisms present, and it should not be assumed that these will necessarily be the same in different populations of the same species. These differences are due, at least in part, to the existence of different mutations conferring target site resistance. Consequently, selection for different mutations may result in different patterns of cross-resistance. Enhanced metabolism can affect even closely related herbicides to differing degrees. For example, populations of *Alopecurus myosuroides* (blackgrass) with an enhanced metabolism mechanism show resistance to pendimethalin but not to trifluralin, despite both being dinitroanilines. This is due to differences in the vulnerability of these two herbicides to oxidative metabolism. Consequently, care is needed when trying to predict the efficacy of alternative herbicides.

Mixtures and Sequences

The use of two or more herbicides which have differing modes of action can reduce the selection for resistant genotypes. Ideally, each component in a mixture should:

- Be active at different target sites
- Have a high level of efficacy
- Be detoxified by different biochemical pathways
- Have similar persistence in the soil (if it is a residual herbicide)
- Exert negative cross-resistance
- Synergise the activity of the other component

No mixture is likely to have all these attributes, but the first two listed are the most important. There is a risk that mixtures will select for resistance to both components in the longer term. One practical advantage of sequences of two herbicides compared with mixtures is that a better appraisal of the efficacy of each herbicide component is possible, provided that sufficient time elapses between each application. A disadvantage with sequences is that two separate applications have to be made and it is possible that the later application will be less effective on weeds surviving the first application. If these are resistant, then the second herbicide in the sequence may increase selection for resistant individuals by killing the susceptible plants which were damaged but not killed by the first application, but allowing the larger, less affected, resistant plants to survive. This has been cited as one reason why ALS-resistant *Stellaria media* has evolved in Scotland recently (2000), despite the regular use of a sequence incorporating mecoprop, a herbicide with a different mode of action.

Herbicide Rotations

Rotation of herbicides from different chemical groups in successive years should reduce selection for resistance. This is a key element in most resistance prevention programmes. The value of this approach depends on the extent of cross-resistance, and whether multiple resistance occurs owing to the presence of several different resistance mechanisms. A practical problem can be the lack of awareness by farmers of the different groups of herbicides that exist. In Australia a scheme has been introduced in which identifying letters are included on the product label as a means of enabling farmers to distinguish products with different modes of action.

Farming Practices and Resistance: A Case Study

Herbicide resistance became a critical problem in Australian agriculture, after many Australian sheep farmers began to exclusively grow wheat in their pastures in the 1970s. Introduced varieties of ryegrass, while good for grazing sheep, compete intensely with wheat. Ryegrasses produce so many seeds that, if left unchecked, they can completely choke a field. Herbicides provided excellent control, while reducing soil disrupting because of less need to plough. Within little more than a decade, ryegrass and other weeds began to develop resistance. In response Australian farmers changed methods. By 1983, patches of ryegrass had become immune to Hoegrass, a family of herbicides that inhibit an enzyme called acetyl coenzyme A carboxylase.

Ryegrass populations were large, and had substantial genetic diversity, because farmers had planted many varieties. Ryegrass is cross-pollinated by wind, so genes shuffle frequently. To control its distribution farmers sprayed inexpensive Hoegrass, creating selection pressure. In addition, farmers sometimes diluted the herbicide in order to save money, which allowed some plants to survive application. When resistance appeared farmers turned to a group of herbicides that block acetolactate synthase. Once again, ryegrass in Australia evolved a kind of "cross-resistance" that allowed it to rapidly break down a variety of herbicides. Four classes of herbicides become ineffective within a few years. In 2013 only two herbicide classes, called Photosystem II and long-chain fatty acid inhibitors, were effective against ryegrass.

List of Common Herbicides

Synthetic Herbicides

- 2,4-D is a broadleaf herbicide in the phenoxy group used in turf and no-till field crop production. Now, it is mainly used in a blend with other herbicides to allow lower rates of herbicides to be used; it is the most widely used herbicide in the world, and third most commonly used in the United States. It is an example of synthetic auxin (plant hormone).

- Aminopyralid is a broadleaf herbicide in the pyridine group, used to control weeds on grassland, such as docks, thistles and nettles. It is notorious for its ability to persist in compost.

- Atrazine, a triazine herbicide, is used in corn and sorghum for control of broadleaf weeds and grasses. Still used because of its low cost and because it works well on a broad spectrum of weeds common in the US corn belt, atrazine is commonly used with other herbicides to reduce the overall rate of atrazine and to lower the potential for groundwater contamination; it is a photosystem II inhibitor.

- Clopyralid is a broadleaf herbicide in the pyridine group, used mainly in turf, rangeland, and for control of noxious thistles. Notorious for its ability to persist in compost, it is another example of synthetic auxin.

- Dicamba, a postemergent broadleaf herbicide with some soil activity, is used on turf and field corn. It is another example of a synthetic auxin.

- Glufosinate ammonium, a broad-spectrum contact herbicide, is used to control weeds after the crop emerges or for total vegetation control on land not used for cultivation.

- Fluazifop (Fuselade Forte), a post emergence, foliar absorbed, translocated grass-selective herbicide with little residual action. It is used on a very wide range of broad leaved crops for control of annual and perennial grasses.

- Fluroxypyr, a systemic, selective herbicide, is used for the control of broad-leaved weeds in small grain cereals, maize, pastures, rangeland and turf. It is a synthetic auxin. In cereal growing, fluroxypyr's key importance is control of cleavers, *Galium aparine*. Other key broadleaf weeds are also controlled.

- Glyphosate, a systemic nonselective herbicide, is used in no-till burndown and for weed control in crops genetically modified to resist its effects. It is an example of an EPSPs inhibitor.

- Imazapyr a nonselective herbicide, is used for the control of a broad range of weeds, including terrestrial annual and perennial grasses and broadleaf herbs, woody species, and riparian and emergent aquatic species.

- Imazapic, a selective herbicide for both the pre- and postemergent control of some annual and perennial grasses and some broadleaf weeds, kills plants by inhibiting the production of branched chain amino acids (valine, leucine, and isoleucine), which are necessary for protein synthesis and cell growth.

- Imazamox, an imidazolinone manufactured by BASF for postemergence application that is an acetolactate synthase (ALS) inhibitor. Sold under trade names Raptor, Beyond, and Clearcast.

- Linuron is a nonselective herbicide used in the control of grasses and broadleaf weeds. It works by inhibiting photosynthesis.

- MCPA (2-methyl-4-chlorophenoxyacetic acid) is a phenoxy herbicide selective for broadleaf plants and widely used in cereals and pasture.

- Metolachlor is a pre-emergent herbicide widely used for control of annual grasses in corn and sorghum; it has displaced some of the atrazine in these uses.

- Paraquat is a nonselective contact herbicide used for no-till burndown and in aerial destruction of marijuana and coca plantings. It is more acutely toxic to people than any other herbicide in widespread commercial use.

- Pendimethalin, a pre-emergent herbicide, is widely used to control annual grasses and some broad-leaf weeds in a wide range of crops, including corn, soybeans, wheat, cotton, many tree and vine crops, and many turfgrass species.

- Picloram, a pyridine herbicide, mainly is used to control unwanted trees in pastures and edges of fields. It is another synthetic auxin.

- Sodium chlorate *(disused/banned in some countries)*, a nonselective herbicide, is considered phytotoxic to all green plant parts. It can also kill through root absorption.

- Triclopyr, a systemic, foliar herbicide in the pyridine group, is used to control broadleaf weeds while leaving grasses and conifers unaffected.

- Several sulfonylureas, including Flazasulfuron and Metsulfuron-methyl, which act as ALS inhibitors and in some cases are taken up from the soil via the roots.

Organic Herbicides

Recently, the term "organic" has come to imply products used in organic farming. Under this definition, an organic herbicide is one that can be used in a farming enterprise that has been classified as organic. Commercially sold organic herbicides are expensive and may not be affordable for commercial farming. Depending on the application, they may be less effective than synthetic herbicides and are generally used along with cultural and mechanical weed control practices.

Homemade organic herbicides include:

- Corn gluten meal (CGM) is a natural pre-emergence weed control used in turfgrass, which reduces germination of many broadleaf and grass weeds.

- Vinegar is effective for 5–20% solutions of acetic acid, with higher concentrations most effective, but it mainly destroys surface growth, so respraying to treat regrowth is needed. Resistant plants generally succumb when weakened by respraying.

- Steam has been applied commercially, but is now considered uneconomical and inadequate. It controls surface growth but not underground growth and so respraying to treat regrowth of perennials is needed.

- Flame is considered more effective than steam, but suffers from the same difficulties.

- D-limonene (citrus oil) is a natural degreasing agent that strips the waxy skin or cuticle from weeds, causing dehydration and ultimately death.

- Saltwater or salt applied in appropriate strengths to the rootzone will kill most plants.

- Monocerin produced by certain fungi will kill certain weeds such as Johnson grass.

Of Historical interest and Other

- 2,4,5-Trichlorophenoxyacetic acid (2,4,5-T) was a widely used broadleaf herbicide until being phased out starting in the late 1970s. While 2,4,5-T itself is of only moderate toxicity, the manufacturing process for 2,4,5-T contaminates this chemical with trace amounts of 2,3,7,8-tetrachlorodibenzo-p-dioxin (TCDD). TCDD is extremely toxic to humans. With proper temperature control during production of 2,4,5-T, TCDD levels can be held to about .005 ppm. Before the TCDD risk was well understood, early production facilities lacked proper temperature controls. Individual batches tested later were found to have as much as 60 ppm of TCDD. 2,4,5-T was withdrawn from use in the USA in 1983, at a time of heightened public sensitivity about chemical hazards in the environment. Public concern about dioxins was high, and production and use of other (non-herbicide) chemicals potentially containing TCDD contamination was also withdrawn. These included pentachlorophenol (a wood preservative) and PCBs (mainly used as stabilizing agents in transformer oil). Some feel that the 2,4,5-T withdrawal was not based on sound science. 2,4,5-T has since largely been replaced by dicamba and triclopyr.

- Agent Orange was a herbicide blend used by the British military during the Malayan Emergency and the U.S. military during the Vietnam War between January 1965 and April 1970 as a defoliant. It was a 50/50 mixture of the *n*-butyl esters of 2,4,5-T and 2,4-D. Because of TCDD contamination in the 2,4,5-T component, it has been blamed for serious illnesses in many people who were exposed to it. However, research on populations exposed to its dioxin contaminant have been inconsistent and inconclusive.

- Diesel, and other heavy oil derivatives, are known to be informally used at times, but are usually banned for this purpose.

Molluscicide

Molluscicides, also known as snail baits and snail pellets, are pesticides against molluscs, which are usually used in agriculture or gardening, in order to control gastropod pests specifically slugs and snails which damage crops or other valued plants by feeding on them.

A number of chemicals can be employed as a molluscicide:

- Metal salts such as iron(III) phosphate and aluminium sulfate, relatively non-toxic, also used in organic gardening

- Metaldehyde

- Methiocarb

- Acetylcholinesterase inhibitors, highly toxic to other animals and humans, acts also as a contact poison

Accidental Poisonings

Metal salt-based molluscicides are not toxic to higher animals. However, metaldehyde-based and especially acetylcholinesterase inhibitor-based products are highly toxic, and have resulted in many deaths of pets and humans. Some products contain a bittering agent that reduces but does not eliminate the risk of accidental poisoning. Anticholinergic drugs such as atropine can be used as an antidote for acetylcholinesterase inhibitor poisoning. There is no antidote for metaldehyde, the treatment is symptomatic.

Methiocarb can cause acute toxicity to people exposed to it for long periods of time and will also poison water organisms.

Rodenticide

Rodenticides, colloquially rat poison, are typically non-specific pest control chemicals made and sold for the purpose of killing rodents.

Typical rat poison bait station (Germany, 2010

Some rodenticides are lethal after one exposure while others require more than one. Rodents are disinclined to gorge on an unknown food (perhaps reflecting an adaptation to their inability to vomit), preferring to sample, wait and observe whether it makes them or other rats sick. This phenomenon of bait shyness or poison shyness is the rationale for poisons that kill only after multiple doses.

Besides being directly toxic to the mammals that ingest them, including dogs, cats, and humans, many rodenticides present a secondary poisoning risk to animals that hunt or scavenge the dead corpses of rats.

Chemical Preparations

Anticoagulants

Anticoagulants are defined as chronic (death occurs one to two weeks after ingestion of the lethal dose, rarely sooner), single-dose (second generation) or multiple-dose (first generation) rodenticides, acting by effective blocking of the vitamin K cycle, resulting in inability to produce essential blood-clotting factors — mainly coagulation factors II (prothrombin) and VII (proconvertin).

In addition to this specific metabolic disruption, massive toxic doses of 4-hydroxycoumarin, 4-thiochromenone and indandione anticoagulants cause damage to tiny blood vessels (capillaries), increasing their permeability, causing diffuse internal bleeding. These effects are gradual, developing over several days. In the final phase of the intoxication, the exhausted rodent collapses due to hemorrhagic shock or severe anemia and dies calmly. The question of whether the use of these rodenticides can be considered humane has been raised.

The main benefit of anticoagulants over other poisons is that the time taken for the poison to induce death means that the rats do not associate the damage with their feeding habits.

- First generation rodenticidal anticoagulants generally have shorter elimination half-lives, require higher concentrations (usually between 0.005% and 0.1%) and consecutive intake over days in order to accumulate the lethal dose, and are less toxic than second generation agents.

- Second generation agents are far more toxic than first generation. They are generally applied in lower concentrations in baits — usually on the order of 0.001% to 0.005% — are lethal after a single ingestion of bait and are also effective against strains of rodents that became resistant to first generation anticoagulants; thus, the second generation anticoagulants are sometimes referred to as "superwarfarins".

Class	Examples
Coumarins/4-hydroxycoumarins	• First generation: warfarin, coumatetralyl • Second generation: difenacoum, brodifacoum, flocoumafen and bromadiolone.
1,3-indandiones	diphacinone, chlorophacinone, pindone These are harder to group by generation. According to some sources, the indandiones are considered second generation. However, according to the U.S. Environmental Protection Agency, examples of first generation agents include chlorophacinone and diphacinone.
4-thiochromenones	Difethialone is considered a second generation anticoagulant rodenticide .
Indirect	Sometimes, anticoagulant rodenticides are potentiated by an antibiotic or bacteriostatic agent, most commonly sulfaquinoxaline. The aim of this association is that the antibiotic suppresses intestinal symbiotic microflora, which are a source of vitamin K. Diminished production of vitamin K by the intestinal microflora contributes to the action of anticoagulants. Added vitamin D also has a synergistic effect with anticoagulants.

Vitamin K_1 has been suggested, and successfully used, as antidote for pets or humans accidentally or intentionally exposed to anticoagulant poisons. Some of these poisons act by inhibiting liver functions and in advanced stages of poisoning, several blood-clotting factors are absent, and the volume of circulating blood is diminished, so that a blood transfusion (optionally with the clotting factors present) can save a person who has been poisoned, an advantage over some older poisons.

Metal Phosphides

Rat poison vendor's stall at a market in Linxia City, China

Metal phosphides have been used as a means of killing rodents and are considered single-dose fast acting rodenticides (death occurs commonly within 1–3 days after single bait ingestion). A bait consisting of food and a phosphide (usually zinc phosphide) is left where the rodents can eat it. The acid in the digestive system of the rodent reacts with the phosphide to generate the toxic phosphine gas. This method of vermin control has possible use in places where rodents are resistant to some of the anticoagulants, particularly for control of house and field mice; zinc phosphide baits are also cheaper than most second-generation anticoagulants, so that sometimes, in the case of large infestation by rodents, their population is initially reduced by copious amounts of zinc phosphide bait applied, and the rest of population that survived the initial fast-acting poison is then eradicated by prolonged feeding on anticoagulant bait. Inversely, the individual rodents, that survived anticoagulant bait poisoning (rest population) can be eradicated by pre-baiting them with nontoxic bait for a week or two (this is important to overcome bait shyness, and to get rodents used to feeding in specific areas by specific food, especially in eradicating rats) and subsequently applying poisoned bait of the same sort as used for pre-baiting until all consumption of the bait ceases (usually within 2–4 days). These methods of alternating rodenticides with different modes of action gives actual or almost 100% eradications of the rodent population in the area, if the acceptance/palatability of baits are good (i.e., rodents feed on it readily).

Zinc phosphide is typically added to rodent baits in a concentration of 0.75% to 2.0%. The baits have strong, pungent garlic-like odor due to the phosphine liberated by hydrolysis. The odor attracts (or, at least, does not repel) rodents, but has an repulsive effect on other mammals. Birds, notably wild turkeys, are not sensitive to the smell, and will feed on the bait, and thus become collateral damage.

The tablets or pellets (usually aluminium, calcium or magnesium phosphide for fumigation/gassing) may also contain other chemicals which evolve ammonia, which helps to reduce the potential for spontaneous combustion or explosion of the phosphine gas.

Metal phosphides do not accumulate in the tissues of poisoned animals, so the risk of secondary poisoning is low.

Before the advent of anticoagulants, phosphides were the favored kind of rat poison. During World War II, they came into use in United States because of shortage of strychnine due to the Japanese occupation of the territories where the strychnine tree is grown. Phosphides are rather fast-acting rat poisons, resulting in the rats dying usually in open areas, instead of in the affected buildings.

Phosphides used as rodenticides include:

- aluminium phosphide (fumigant only)

- calcium phosphide (fumigant only)

- magnesium phosphide (fumigant only)

- zinc phosphide (bait only)

Hypercalcemia

Calciferols (vitamins D), cholecalciferol (vitamin D_3) and ergocalciferol (vitamin D_2) are used as rodenticides. They are toxic to rodents for the same reason they are important to humans: they affect calcium and phosphate homeostasis in the body. Vitamins D are essential in minute quantities (few IUs per kilogram body weight daily, only a fraction of a milligram), and like most fat soluble vitamins, they are toxic in larger doses, causing hypervitaminosis. If the poisoning is severe enough (that is, if the dose of the toxin is high enough), it leads to death. In rodents that consume the rodenticidal bait, it causes hypercalcemia, raising the calcium level, mainly by increasing calcium absorption from food, mobilising bone-matrix-fixed calcium into ionised form (mainly monohydrogencarbonate calcium cation, partially bound to plasma proteins, $[CaHCO_3]^+$), which circulates dissolved in the blood plasma. After ingestion of a lethal dose, the free calcium levels are raised sufficiently that blood vessels, kidneys, the stomach wall and lungs are mineralised/calcificated (formation of calcificates, crystals of calcium salts/complexes in the tissues, damaging them), leading further to heart problems (myocardial tissue is sensitive to variations of free calcium levels, affecting both myocardial contractibility and excitation propagation between atrias and ventriculas), bleeding (due to capillary damage) and possibly kidney failure. It is considered to be single-dose, cumulative (depending on concentration used; the common 0.075% bait concentration is lethal to most rodents after a single intake of larger portions of the bait) or sub-chronic (death occurring usually within days to one week after ingestion of the bait). Applied concentrations are 0.075% cholecalciferol and 0.1% ergocalciferol when used alone, wihich can kill a rodent or a rat.

There is an important feature of calciferols toxicology, that they are synergistic with anticoagulant toxicants, that means, that mixtures of anticoagulants and calciferols in same bait are more toxic than a sum of toxicities of the anticoagulant and the calciferol in the bait, so that a massive hypercalcemic effect can be achieved by a substantially lower calciferol content in the bait, and vice versa, a more pronounced anticoagulant/hemorrhagic effects are observed if the calciferol is present. This synergism is mostly used in calciferol low concentration baits, because effective concentrations of calciferols are more expensive than effective concentrations of most anticoagulants.

The first application of a calciferol in rodenticidal bait was in the Sorex product Sorexa D (with a different formula than today's Sorexa D), back in the early 1970s, which contained 0.025% warfarin and 0.1% ergocalciferol. Today, Sorexa CD contains a 0.0025% difenacoum and 0.075% cholecalciferol combination. Numerous other brand products containing either 0.075-0.1% calciferols (e.g. Quintox) alone or alongside an anticoagulant are marketed.

The Merck Veterinary Manual states the following:

Although this rodenticide [cholecalciferol] was introduced with claims that it was less toxic to nontarget species than to rodents, clinical experience has shown that rodenticides containing cholecalciferol are a significant health threat to dogs and cats. Cholecalciferol produces hypercalcemia, which results in systemic calcification of soft tissue, leading to renal failure, cardiac abnormalities, hypertension, CNS depression and GI upset. Signs generally develop within 18-36 hours of ingestion and can include depression, anorexia, polyuria and polydipsia. As serum calcium concentrations increase, clinical signs become more severe. ... GI smooth muscle excitability decreases and is manifest by anorexia, vomiting and constipation. ... Loss of renal concentrating ability is a direct result of hypercalcemia. As hypercalcemia persists, mineralization of the kidneys results in progressive renal insufficiency."

Additional anticoagulant renders the bait more toxic to pets as well as human. Upon single ingestion, solely calciferol-based baits are considered generally safer to birds than second generation anticoagulants or acute toxicants. A specific antidote for calciferol intoxication is calcitonin, a hormone that lowers the blood levels of calcium. The therapy with commercially available calcitonin preparations is, however, expensive.

Other

Civilian Public Service worker distributes rat poison for typhus control in Gulfport, Mississippi, ca. 1945.

Other chemical poisons include:

- ANTU (α-naphthylthiourea; specific against Brown rat, *Rattus norvegicus*)
- Arsenic trioxide
- Barium carbonate
- Chloralose (a narcotic prodrug)
- Crimidine (inhibits metabolism of vitamin B_6)
- 1,3-Difluoro-2-propanol ("Gliftor")
- Endrin (organochlorine insecticide, used in the past for extermination of voles in fields)
- Fluoroacetamide ("1081")
- Phosacetim (a delayed-action organophosphate)
- White phosphorus
- Pyrinuron (an urea derivative)
- Scilliroside and other cardiac glycosides like oleandrin and digoxin
- Sodium fluoroacetate ("1080")
- Strychnine (A naturally occurring convulsant and stimulant)
- Tetramethylenedisulfotetramine ("tetramine")
- Thallium sulfate
- Nitrophenols like bromethalin and 2,4-dinitrophenol (cause high fever and brain swelling, no known antidote)

- Zyklon B/Uragan D2 (hydrogen cyanide gas absorbed in an inert carrier)

Combinations

In some countries, fixed three-component rodenticides, i.e., anticoagulant + antibiotic + vitamin D, are used. Associations of a second-generation anticoagulant with an antibiotic and/or vitamin D are considered to be effective even against most resistant strains of rodents, though some second generation anticoagulants (namely brodifacoum and difethialone), in bait concentrations of 0.0025% to 0.005% are so toxic that resistance is unknown, and even rodents resistant to other rodenticides are reliably exterminated by application of these most toxic anticoagulants.

Alternatives

More environmentally-safe preparations, such as powdered corn cob, have been developed and were approved in the EU and patented in the US in 2013. These preparations rely on dehydration to cause death.

Non-target Issues

Secondary Poisoning and Risks to Wildlife

One of the potential problems when using rodenticides is that dead or weakened rodents may be eaten by other wildlife, either predators or scavengers. Members of the public deploying rodenticides may not be aware of this or may not follow the product's instructions closely enough.

The faster a rodenticide acts, the more critical this problem may be. For the fast-acting rodenticide bromethalin, for example, there is no diagnostic test or antidote.

This has led environmental researchers to conclude that low strength, long duration rodenticides (generally first generation anticoagulants) are the best balance between maximum effect and minimum risk.

Proposed US Legislation Change

In 2008, after assessing human health and ecological effects, as well as benefits, the US Environmental Protection Agency (EPA) announced measures to reduce risks associated with ten rodenticides. New restrictions by sale and distribution restrictions, minimum package size requirements, use site restriction, and tamper resistant products would have taken effect in 2011. The regulations were delayed pending a legal challenge by manufacturer Reckitt-Benkiser.

Notable Rat Eradications

The entire rat populations of several islands have been eradicated, most notably Campbell Island, New Zealand (11,300 ha), Hawadax Island, Alaska (formerly known as Rat Island, 2,670 ha) and Canna, Scotland (1,030 ha, declared rat-free in 2008).

Alberta, Canada, through a combination of climate and control, is believed to be rat-free.

Toxicity Label

Toxicity labels viz; red label, yellow label, blue label and green label are mandatory labels employed on pesticide containers in India identifying the level of toxicity (that is, the toxicity class) of the contained pesticide. The schemes follows from the *Insecticides Act* of 1968 and the *Insecticides Rules* of 1971.

The labeling follows a general scheme as laid down in the *Insecticides Rules, 1971*, and contains information such as brand name, name of manufacturer, name of the antidote in case of accidental consumption etc. A major aspect of the label is a color mark which represents the toxicity of the material by a color code. Thus the labelling scheme proposes four different colour labels: viz red, yellow, blue, and green.

Label	Name	Level of toxicity	Oral lethal dose mg per kg body weight of test animal	Listed chemicals
POISON	Red label	Extremely toxic	1-50	Monocrotophos, zinc phosphide, ethyl mercury acetate, and others.
POISON	Yellow label	Highly toxic	51-500	Endosulfan, carbaryl, quinalphos, and others.
DANGER KEEP OUT OF THE REACH OF CHILDREN	Blue label	Moderately toxic	501-5000	Malathion, thiram, glyphosate, and others.
CAUTION	Green label	Slightly toxic	More than 5000	Mancozeb, oxyfluorfen, mosquito repellant oils and liquids, and most other household insecticides.

The toxicity classification applies only to pesticides which are allowed to be sold in India. Some of the classified pesticides may be banned in somes states of India, by decision of the state governments. Some of the red-label and yellow-label pesticides were banned in the state of Kerala following the Endosulfan protests of 2011.

References

- Campbell, Malcolm (2003-09-19). "Fact Sheet: Milk Fungicide". Australian Broadcasting Corporation. Retrieved 2009-04-01.

- Quastel, J. H. (1950). "2,4-Dichlorophenoxyacetic Acid (2,4-D) as a Selective Herbicide". Agricultural Control Chemicals. Advances in Chemistry. 1. p. 244. doi:10.1021/ba-1950-0001.ch045. ISBN 0-8412-2442-0.

- Shaner, D. L.; Leonard, P. (2001). "Regulatory aspects of resistance management for herbicides and other crop protection products". In Powles, S. B.; Shaner, D. L. Herbicide Resistance and World Grains. CRC Press, Boca Raton, FL. pp. 279–294. ISBN 9781420039085.

- Smith (18 July 1995). "8: Fate of herbicides in the environment". Handbook of Weed Management Systems. CRC Press. pp. 245–278. ISBN 978-0-8247-9547-4.

- Powles, S. B.; Shaner, D. L., eds. (2001). Herbicide Resistance and World Grains. CRC Press, Boca Raton, FL. p. 328. ISBN 9781420039085.

- Moss, S. R. (2002). "Herbicide-Resistant Weeds". In Naylor,, R. E. L. Weed management handbook (9th ed.). Blackwell Science Ltd. pp. 225–252. ISBN 0-632-05732-7.

- "EU approves powdered corn cob as biocidal active". Chemical Watch: Global Risk & Regulation News. 15 August 2013. Retrieved 22 August 2013.

- "Island which spent £600,000 getting rid of rats over-run with rabbits". Daily Telegraph. 27 April 2010. Retrieved 4 April 2015.

4

Various Aspects of Pesticides

Pesticide use is a hotly debated topic due to its potential hazards to the environment, people and other organisms. The information presented in this section draws on these issues and discusses subjects like pesticide resistance, restricted use pesticide, non-pesticide management, pesticide regulation in the United States and pesticide degradation. This chapter elucidates the crucial theories and methods of pesticide use.

Pesticide Resistance

Pesticide resistance describes the decreased susceptibility of a pest population to a pesticide that was previously effective at controlling the pest. Pest species evolve pesticide resistance via natural selection: the most resistant specimens survive and pass on their genetic traits to their offspring.

The Insecticide Resistance Action Committee (IRAC) definition of insecticide resistance is 'a heritable change in the sensitivity of a pest population that is reflected in the repeated failure of a product to achieve the expected level of control when used according to the label recommendation for that pest species'.

Pesticide resistance is increasing. Farmers in the USA lost 7% of their crops to pests in the 1940s; over the 1980s and 1990s, the loss was 13%, even though more pesticides were being used. Over 500 species of pests have evolved a resistance to a pesticide. Other sources estimate the number to be around 1000 species since 1945.

Although the evolution of pesticide resistance is usually discussed as a result of pesticide use, it is important to keep in mind that pest populations can also adapt to non-chemical methods of control. For example, the northern corn rootworm (*Diabrotica barberi*) became adapted to a corn-soybean crop rotation by spending the year when field is planted to soybeans in a diapause.

As of 2014 few new weed killers are near commercialization, and none with a novel, resistance-free mode of action.

Causes

Pesticide resistance probably stems from multiple factors:

Many pest species produce large broods. This increases the probability of mutations and ensures the rapid expansion of resistant populations.

Pest species had been exposed to natural toxins long before agriculture began. For example, many plants produce phytotoxins to protect them from herbivores. As a result, coevolution of herbivores and their host plants required development of the physiological capability to detoxify or tolerate poisons.

Humans often rely almost exclusively on pesticides for pest control. This increases selection pressure towards resistance. Pesticides that fail to break down quickly contribute to selection for resistant strains even after they are no longer being applied.

In response to resistance, managers may increase pesticide quantities/frequency, which exacerbates the problem. In addition, some pesticides are toxic toward species that feed on or compete with pests. This can allow the pest population to expand, requiring more pesticides. This is sometimes referred to as *pesticide trap*, or a *pesticide treadmill*, since farmers progressively pay more for less benefit.

Insect predators and parasites generally have smaller populations and are less likely to evolve resistance than are pesticides' primary targets, such as mosquitoes and those that feed on plants. Weakening them allows the pests to flourish. Alternatively, resistant predators can be bred in laboratories.

Pests with limited diets are more likely to evolve resistance, because they are exposed to higher pesticide concentrations and has less opportunity to breed with unexposed populations.

Pests with shorter generation times develop resistance more quickly than others.

Examples

Resistance has evolved in multiple species: Resistance to insecticides was first documented by A. L. Melander in 1914 when scale insects demonstrated resistance to an inorganic insecticide. Between 1914 and 1946, 11 additional cases were recorded. The development of organic insecticides, such as DDT, gave hope that insecticide resistance was a dead issue. However, by 1947 housefly resistance to DDT had evolved. With the introduction of every new insecticide class – cyclodienes, carbamates, formamidines, organophosphates, pyrethroids, even *Bacillus thuringiensis* – cases of resistance surfaced within two to 20 years.

- In the US, studies have shown that fruit flies that infest orange groves were becoming resistant to malathion.

- In Hawaii, Japan and Tennessee, the diamondback moth evolved a resistance to *Bacillus thuringiensis* about three years after it began to be used heavily.

- In England, rats in certain areas have evolved resistance that allows them to consume up to five times as much rat poison as normal rats without dying.

- DDT is no longer effective in preventing malaria in some places.

- In the southern United States, *Amaranthus palmeri*, which interferes with cotton production, has evolved resistance to the herbicide glyphosate.

- The Colorado potato beetle has evolved resistance to 52 different compounds belonging to all major insecticide classes. Resistance levels vary across populations and between beetle life stages, but in some cases can be very high (up to 2,000-fold).

Multiple and Cross-resistance

Multiple resistance pests are resistant to more than one class of pesticide. This can happen when pesticides are used in sequence, with a new class replacing one to which pests display resistance with another. *Cross resistance*, a related phenomenon, occurs when the genetic mutation that made the pest resistant to one pesticide also makes it resistant to others, often those with a similar mechanism of action.

Adaptation

Pests becomes resistant by evolving physiological changes that protect them from the chemical.

One protection mechanism is to increase the number of copies of a gene, allowing the organism to produce more of a protective enzyme that breaks the pesticide into less toxic chemicals. Such enzymes include esterases, glutathione transferases, and mixed microsomal oxidases.

Alternatively, the number and/or sensitivity of biochemical receptors that bind to the pesticide may be reduced.

Behavioral resistance has been described for some chemicals. For example, some Anopheles mosquitoes evolved a preference for resting outside that kept them away from pesticide sprayed on interior walls.

Resistance may involve rapid excretion of toxins, secretion of them within the body away from vulnerable tissues and decreased penetration through the body wall.

Mutation in only a single gene can lead to the evolution of a resistant organism. In other cases, multiple genes are involved. Resistant genes are usually autosomal. This means that they are located on autosomes (as opposed to chromosomes). As a result, resistance is inherited similarly in males and females. Also, resistance is usually inherited as an incompletely dominant trait. When a resistant individual mates with a susceptible individual, their progeny generally has a level of resistance intermediate between the parents.

Adaptation to pesticides comes with an evolutionary cost, usually decreasing relative fitness of organisms in the absence of pesticides. Resistant individuals often have reduced reproductive output, life expectancy, mobility, etc. Non-resistant individuals grow in frequency in the absence of pesticides, offering one way to combat resistance.

Blowfly maggots produce an enzyme that confers resistance to organochloride insecticides. Scientists have researched ways to use this enzyme to break down pesticides in the environment, which would detoxify them and prevent harmful environmental effects. A similar enzyme produced by soil bacteria that also breaks down organochlorides works faster and remains stable in a variety of conditions.

Management

Integrated pest management (IPM) approach provides a balanced approach to minimizing resistance.

Resistance can be managed by reducing use of a pesticide. This allows non-resistant organisms to out-compete resistant strains. They can later be killed by returning to use of the pesticide.

A complementary approach is to site untreated refuges near treated croplands where susceptible pests can survive.

When pesticides are the sole or predominant method of pest control, resistance is commonly managed through pesticide rotation. This involves switching among pesticide classes with different modes of action to delay or mitigate pest resistance. The U.S. Environmental Protection Agency (EPA) designates different classes of fungicides, herbicides and insecticides. Manufacturers may recommend no more than a specified number of consecutive applications of a pesticide class be made before moving to a different pesticide class.

Two or more pesticides with different modes of action can be tankmixed on the farm to improve results and delay or mitigate existing pest resistance.

Status

Glyphosate

Glyphosate-resistant weeds are now present in the vast majority of soybean, cotton, and corn farms in some U.S. states. Weeds resistant to multiple herbicide modes of action are also on the rise.

Before glyphosate, most herbicides would kill a limited number of weed species, forcing farmers to continually rotate their crops and herbicides to prevent resistance. Glyphosate disrupts the ability of most plants to construct new proteins. Glyphosate-tolerant transgenic crops are not affected.

A weed family that includes waterhemp (Amaranthus rudis) has developed glyphosate-resistant strains. A 2008 to 2009 survey of 144 populations of waterhemp in 41 Missouri counties revealed glyphosate resistance in 69%. Weed surveys from some 500 sites throughout Iowa in 2011 and 2012 revealed glyphosate resistance in approximately 64% of waterhemp samples.

In response to the rise in glyphosate resistance, farmers turned to other herbicides—applying several in a single season. In the United States, most midwestern and southern farmers continue to use glyphosate because it still controls most weed species, applying other herbicides, known as residuals, to deal with resistance.

The use of multiple herbicides appears to have slowed the spread of glyphosate resistance. From 2005 through 2010 researchers discovered 13 different weed species that had developed resistance to glyphosate. From 2010-2014 only two more were discovered.

A 2013 Missouri survey showed that multiply-resistant weeds had spread. 43% of the sampled weed populations were resistant to two different herbicides, 6% to three and 0.5% to four. In Iowa a survey revealed dual resistance in 89% of waterhemp populations, 25% resistant to three and 10% resistant to five.

Resistance increases pesticide costs. For southern cotton, herbicide costs climbed from between $50 and $75 per hectare a few years ago to about $370 per hectare in 2014. In the South, resistance contributed to the shift that reduced cotton planting by 70% in Arkansas and 60% in Tennessee. For soybeans in Illinois costs rose from about $25 to $160 per hectare.

B. Thuringiensis

During 2009 and 2010, some Iowa fields showed severe injury to corn producing Bt toxin Cry3Bb1 by western corn rootworm. During 2011, mCry3A corn also displayed insect damage, including cross-resistance between these toxins. Resistance persisted and spread in Iowa. Bt corn that targets western corn rootworm does not produce a high dose of Bt toxin, and displays less resistance than that seen in a high-dose Bt crop.

Products such as Capture LFR (containing the pyrethroid Bifenthrin) and SmartChoice (containing a pyrethroid and an organophosphate) have been increasingly used to complement Bt crops that farmers find alone to be unable to prevent insect-driven injury. Multiple studies have found the practice to be either ineffective or to accelerate the development of resistant strains.

Restricted Use Pesticide

Restricted use pesticides or "RUP" are pesticides not available to the general public in the United States. The "Restricted Use" classification restricts a product, or its uses, to use by a certificated pesticide applicator or under the direct supervision of a certified applicator. This means that a license is required to purchase and apply the product. Certification programs are administered by the federal government, individual states, and by company policies that vary from state to state. This is managed by the United States Environmental Protection Agency (EPA) under the Worker Protection Standard.

Pesticides are classified as "restricted use" for a variety of reasons, such as potential for or history of groundwater contamination. Atrazine is the most widely used restricted-use herbicide. Many insecticides and fungicides used in fruit production are restricted use.

List

The list of restricted use pesticides is maintained by the US EPA.

- Restricted Use Products (RUP) Report

Requirement

The Worker Protection Standard (WPS) identifies the type of requirements that must be satisfied to obtain the proper license needed to purchase and apply restricted use pesticide.

- Worker Protection Standard

The Hazard Communication Standard requires all employers to disclose all hazards to employees separately from the WPS.

Licensed pest control supervisors must maintain application records for 3 years or more, as determined by state and federal laws. These records must identify the date, location, and type of pesticide that has been applied.

Additionally, the licensed pest control supervisors must notify the local government agency that is responsible for air quality to satisfy laws governing the Right to know regarding public health and safety risks when restricted use pesticides are applied outside buildings.

- Right to know

Non-pesticide Management

Non-pesticidal Management (NPM) describes various pest-control techniques which do not rely on pesticides. It is used in organic production of foodstuff, as well as in other situations in which the introduction of toxins is undesirable. Instead of the use of synthetic toxins, pest control is achieved by biological means.

Some examples of Non-Pesticidal Management techniques include:

- Introduction of natural predators.

- Use of naturally occurring insecticides, such as Neem tree products, Margosa, Tulsi / Basil Leaf, Citrus Oil, Eucalyptus Oil, Onion, Garlic spray, Essential Oils. These also refer to as Organic Pesticides.

- Use of trap crops which attract the insects away from the fields. The trap crops are regularly checked and pests are manually removed.

- Pest larvae which were killed by viruses can be crushed and sprayed over fields, thus killing the remaining larvae.

- Field sanitation.

- Timely sowing.

- Nutrient management.

- Maintain proper plant population.

- Go for soil solarisation.

- Deep summer ploughing.

Over years insects have withstood natural calamities and survived successfully. Hence they are able to develop resistance to the extremely toxic chemical pesticides insecticides used by farmers. To be successful, farmers should be knowledgeable and able to identify various crop pests, and their natural enemies (farmer's friendly insects). Farmers should recognize different stages of insects and their behavior. The efforts to minimize pests should aim at restoring the natural balance of insects in crop ecosystem but not elimination of the pest.

Principles of NPM

Encouraging Natural Process in Environment

Crop ecosystem should be diverse by growing inter crops, trap crops, border crops in place of mono cropping. Once the insecticide sprays stopped, natural enemies of crop pests gradually establish and exercise control of crop pests, which can be enhanced with botanical extracts like NSKE, chilli garlic extract, cattle dung urine decoction etc.

Herbal repellents Soberbio Veto contributing lot towards natural process of repelling pest and helping environment

Management Skill

Selecting crop based on soil, water resources, climate and local pest problems that occur regularly, crop rotation, adjustment of sowing dates to avoid endemic pests, setting up light and pheromone traps, keeping sticky traps and bird perches.

Local Resources

Using locally available organic amendments for soil improvement and pest control. The extracts are prepared and used as prophylactic or curative measures aimed to restore the disturbed natural balance.

Labor

Shunning chemical inputs and out of shelf products farmer should invest his labor as main investment like regular monitoring of the crop and following methods like deep summer ploughing and shaking plants to dislodge pod borers in pigeonpea.

Community Approach

Pest problem of a farmer not only depends on the crop and management practices of one farmer in isolation. To a large extent it depends on crops cultivated by neighboring farmers and measures taken by them. Hence, identification of activities that necessitate group action and implementing them would be most important aspect in NPM /sustainable agriculture.

Pesticide Regulation in The United States

Pesticide regulation in the United States is primarily a responsibility of the Environmental Protection Agency.

Background

In most countries, pesticides must be approved for sale and use by a government agency. In the United States, the Environmental Protection Agency (EPA) is responsible for regulating pesticides under the Federal Insecticide, Fungicide, and Rodenticide Act (FIFRA) and the Food Quality Protection Act (FQPA). Studies must be conducted to establish the conditions in which the material is safe to use and the effectiveness against the intended pest(s). The EPA regulates pesticides to ensure that these products do not pose adverse effects to humans or the environment. Pesticides produced before November 1984 continue to be reassessed in order to meet the current scientific and regulatory standards. All registered pesticides are reviewed every 15 years to ensure they meet the proper standards. During the registration process, a label is created. The label contains directions for proper use of the material in addition to safety restrictions. Based on acute toxicity, pesticides are assigned to a Toxicity Class.

Some pesticides are considered too hazardous for sale to the general public and are designated restricted use pesticides. Only certified applicators, who have passed an exam, may purchase or supervise the application of restricted use pesticides. Records of sales and use are required to be maintained and may be audited by government agencies charged with the enforcement of pesticide regulations. These records must be made available to employees and state or territorial environmental regulatory agencies.

The EPA regulates pesticides under two main acts, both of which were amended by the Food Quality Protection Act of 1996. In addition to the EPA, the United States Department of Agriculture (USDA) and the United States Food and Drug Administration (FDA) set standards for the level of pesticide residue that is allowed on or in crops The EPA looks at what the potential human health and environmental effects might be associated with the use of the pesticide.

Additionally, the U.S. EPA uses the National Research Council's four-step process for human health risk assessment: (1) Hazard Identification, (2) Dose-Response Assessment, (3) Exposure Assessment, and (4) Risk Characterization.

History

1940s

FIFRA was passed in 1947 and was a collaboration between the federal government and the chemical industry. It resulted from the increase in pesticide production during and after World War II. At that time, the concern about pesticides was related to pesticide efficacy and producer honesty. FIFRA was passed as a "truth in labeling" law. The goal of the legislation was to maintain pesticide standards while allowing new pesticides to enter the market quickly. The 1947 legislation was an example of an "iron triangle," made up of the House Committee on Agriculture, the United States Department of Agriculture (USDA), and the pesticide industry. The legislation required that pesticide formulas be registered with USDA, and that the pesticide labels were accurate. The legislation was not intended to enact an active regulatory system; it was to enable the creation of a stable marketplace. The political climate in the US prevented major changes to the regulation due to the public's relative lack of concern, USDA's administration of the law, and agricultural interests in congress.

1950s

The Delaney House Committee hearings in 1950-51 were the first instance of government hearings concerning pesticide safety. The hearings resulted in two amendments to the Federal Food, Drug and Cosmetic Act: The Pesticides Control Amendment (PCA) and the Food Additives Amendment

(FAA). These two amendments resulted in Food and Drug Administration (FDA) involvement in pesticide regulation. The PCA of 1954 was the first time Congress passed guidance regarding the establishment of safety limits for pesticide residues on food. PCA authorized the FDA to ban pesticides they determined to be unsafe if they were sprayed directly on food. The Food Additives Amendment, which included the Delaney Clause, prohibited the pesticide residues from any carcinogenic pesticides in processed food. In 1959, FIFRA was amended requiring that pesticides be registered.

1960s

In 1962, *The New Yorker* published a series of essays by Rachel Carson, later published as *Silent Spring*, that publicized the negative effects of pesticide use on wildlife. This, along with new evidence that pesticides could have negative impacts on human health, helped spur the creation of the modern environmental movement. Several government agencies, such as the Public Health Service, the Fish and Wildlife Service, and the President's Science Advisory Committee, found evidence that pesticides were negatively affecting human health. There were attempts throughout the 60's to pass legislation reforming FIFRA. The proposals included: moving FIFRA authority from the USDA to the FDA, providing greater public access to pesticide registration data, and mandating better interagency cooperation. None of these proposals gained enough support to pass in both the House and Senate. However, Congress passed an amendment to FIFRA in 1964 allowing the USDA to suspend or cancel a pesticide's registration in order to "prevent an imminent health hazard".

1970s

In 1970, President Richard Nixon created the Environmental Protection Agency (EPA) and shifted control of pesticide regulation from USDA and FDA to the newly created EPA. By this time public awareness of potential human health and environmental health effects had increased. In addition, some members of congress began to express concerns about the adequacy of pesticide regulation. The growing public concern created political pressure for pesticide regulation reform. These changes contributed to an environment that enabled an overhaul of pesticide regulation. FIFRA was amended in 1972 by the Federal Environmental Pesticides Control Act (FEPCA). FEPCA required manufacturers of new pesticides to perform a variety of tests to prove that the pesticide did not have "unreasonable adverse effects" on human health or the environment. The EPA was given the authority to refuse registration to any pesticide it concluded was unsafe. In addition, pesticide registration data was required to be made available to the public after a pesticide had been registered. Pesticides that had been registered prior to 1972 could only be banned after a special review board was convened and determined the pesticide was hazardous. If this occurred, the indemnity clause of FEPCA required the EPA to compensate pesticide manufacturers, distributors, and users for the value of any unused stock they possessed. FEPCA also required EPA to review the registration data from pesticides registered prior to 1972, but did not appropriate funds for the task. The reregistration process was beset by difficulties throughout the 1970s. FEPCA required the reregistration process to be complete by 1976; however EPA did not begin reregistration until the fall of 1975. Shortly after beginning the reregistration process, congressional investigations revealed that the EPA was taking shortcuts that undermined the purpose of reregistration. EPA was confirming the presence of registration data rather than determining whether the data was adequate. Shortly after this was revealed, an investigation into Industrial Bio-Test Laboratories (IBT) found that it had been routinely falsifying tests. IBT was the nation's largest independent laboratory that conducted toxicity tests. The EPA and the Canadian Department of Health and Welfare determined that "only about 10% of the over 2000 IBT studies which had been submitted in support of pesticide registrations were valid." After these two setbacks, EPA suspended its reregistration program in August 1976. EPA did not restart reregistrations until 1978.

1980s

In the early 1980s industry groups attempted to take advantage of the Reagan election to get amendments to FIFRA passed. The pesticide industry was primarily interested in three things: less public access to industry data, longer exclusive use periods, and the right to sue over use of manufacturer data without the manufacturer's permission. Environmental groups prevented the amendments from passing with help from labor groups that had become concerned about the safety of pesticide workers. Until the late 80's short term reauthorizations of FIFRA were passed, but no changes to the law were enacted. In 1986, the Campaign for Pesticide Reform (CPR) and the National Agricultural Chemicals Association (NACA) submitted a bill to congress proposing amendments to FIFRA. CPR was a coalition of environmental, consumer and labor groups. The various groups were persuaded to work together because they were all unhappy with FIFRA. Pesticide manufacturers were frustrated with the amount of time it took to get new pesticides on the market, and wanted an extension on the amount of time they had exclusive right to a pesticide formulation. Environmental groups were frustrated with the indemnity provision of FIFRA, alleging that it made EPA reluctant to ban any pesticides. After a summer of negotiations they submitted a bill that required EPA to review pesticide registration data in order to find data gaps which pesticide manufacturers would have to correct to keep their product on the market. The bill also reinforced the right of the public to access pesticide registration data. However, the CPR-NACA coalition did not include farm groups. This contributed to the failure of the proposal by alienating legislators from farming states. In 1988, congress successfully passed an amendment to FIFRA, and President Reagan signed it into law. It required reregistration of approximately 600 active ingredients within nine years, and required pesticide manufacturers to pay a registration fee to fund the process. The bill repealed EPA's indemnity requirements for manufacturers. The 1988 amendments retained indemnity payments for pesticide users, but the money came from the US Treasury rather than from EPA. It also extended the period of exclusive use for pesticide manufacturers. In 1988 EPA issued an interpretation of the Delaney Clause of 1958 that resolved its contradiction with Pesticides Control Amendment (PCA) of 1954. The PCA required EPA to issue tolerances, or maximum acceptable residue levels, for pesticide residues in food. The Delaney Clause forbade the presence of residue from a carcinogenic pesticide in processed foods, and did not address non-cancer risks. As a result, EPA had different standards for raw and processed foods. EPA's 1988 policy stated that it would use a single standard of negligible risk, regardless of a pesticide's carcinogenicity.

1990s

In 1992, the US Ninth Circuit Court of Appeals ruled that EPA's 1988 policy was invalid, stating that only Congress could change the Delaney Clause. This created incentive for Congress to amend the FFDCA. Another motivating factor was the release in 1993 of the National Academy of Sciences' report "Pesticides in the Diets of Infants and Children." The report made several recommendations for changes in pesticide regulations to protect children's health. First, the report recommended that the EPA stop using cost-benefit analysis to determine whether a pesticide would be registered, and make decisions based solely on health considerations. Second, they recommended that the EPA develop studies to determine the vulnerability of children and infants to pesticides. Third, NAS recommended that EPA add a 10x safety factor as an added protection. Fourth, NAS recommended EPA collect data on the dietary habits of children. Last, they recommended that EPA consider aggregate measurements for pesticides that use the same mechanism of toxicity. After these two events, Congress passed the Food Quality Protection Act (FQPA) in 1996. FQPA amended FFDCA, resolving the Delaney Clause conflict. It mandated a single standard for pesticide residue in food, regardless of the type of food. It also addressed concerns about children's susceptibility to pesticides. It required EPA to assess risk using information about children's eating habits, to consider the children's susceptibility when making decisions about pesticides, and to publish a special determination addressing the safety of each pesticide for children.

Federal Insecticide, Fungicide, and Rodenticide Act

The Federal Insecticide, Fungicide, and Rodenticide Act (FIFRA) act required that all pesticides (whether domestic or foreign) sold or distributed in the United States to be registered. There are four types of registrations under FIFRA for pesticide use

1. Federal Registration Actions: EPA can register pesticides in the United States under Section 3 of FIFRA.

2. Experimental Use Permits (EUPs): EPA can allow manufacturers of pesticides to field test their products under development under Section 5 of FIFRA.

3. Emergency Exemptions: EPA can allow State and Federal agencies, in the event of an emergency pest problem, to permit the use of an unregistered pesticide in a specific area under Section 18 of FIFRA.

4. State-Specific Registration: States can register a new pesticide for general use, or a federally registered product for an additional use, if there is both a demonstrated "special local need," and a tolerance or another clearance through FFDCA under Section 24(c) of FIFRA.

Federal Food, Drug, and Cosmetic Act

The Federal Food, Drug, and Cosmetic Act (FFDCA) requires the EPA to set limits, tolerance levels, on the amount of pesticides that are found on and in food. The tolerance level is the "maximum permissible level for pesticide residues allowed in or on commodities for human food and animal feed."

Food Quality Protection Act of 1996

The EPA must find that a pesticide poses a "reasonable certainty of no harm" before that pesticide can be registered for use on food or feed. Several factors are addressed before a tolerance level is established

- the aggregate, non-occupational exposure from the pesticide (this includes exposure through diet, drinking water, and the use of the pesticide in and around the home);

- the cumulative effects from exposure to different pesticides that produce similar effects in the human body;

- whether there is increased susceptibility to infants and children, or other sensitive subpopulations, from exposure to the pesticide; and

- whether the pesticide produces an effect humans similar to an effect produced by a naturally occurring estrogen or produces other endocrine-disruption effects.

Registration Process

Before a pesticide can be distributed, sold, and used in the United States it must first go through a registration process through the Environmental Protection Agency (EPA). When a pesticide enters the registration process, the EPA considers the "ingredients of the pesticide; the particular site or crop on which it is to be used; the amount, frequency, and timing of its use; and storage and disposal practices." The EPA looks at what the potential human health and environmental effects might be associated with the use of the pesticide. The company that wishes to register the pesticide must provide data from various test that are done using EPA guidelines. These tests include: acute toxicity test (short-term toxicity test) and chronic toxicity test (long-term toxicity test). These tests evaluate: whether the pesticide has the potential to cause adverse effects (including cancer and reproductive system disorders) on humans, wildlife, fish, and plants, including endangered species and non-target organisms; and possible contamination of surface water or ground water from leaching, runoff, and spray drift. The registration process can take upwards of 6 to 9 years, and the cost of registration for a single pesticide is in the range of millions of dollars (Toth, 1996).

The Environmental Protection Agency requires pesticide registrants to report all problems with a registered pesticide. If any problems should arise from any type of pesticide, the EPA takes swift action to recall those products from shelves. These problematic products can be determined as faulty, substandard, or could simply cause injury to the user of the pesticide.

Labeling of Pesticides

A pesticide can only be used legally according to the directions on the label that is included at the time of the sale of the pesticide. The language that is used on the label must be approved by the EPA before it can be sold or distributed in the United States. The purpose of the label is to "provide clear directions for effective product performance while minimizing risk to human health and the environment." A label is a legally binding document that mandates how the pesticide can and must be used and failure to follow the label as written when using the pesticide is a federal offense.

State Regulation

States are authorized to pass their own pesticide regulations provided they are at least as stringent as federal regulations. States receive their pesticide regulation authority through the Federal Insecticide, Fungicide, and Rodenticide Act (FIFRA) and through state pesticide laws. States can require registration of pesticides that are exempt under FIFRA. When there is a special local need for a particular pesticide, states are authorized to add uses to that pesticide under section 24(c) of FIFRA. According to

FIFRA, states are given primary enforcement responsibility when the USEPA has determined that they meet three requirements. First, the state must have state pesticide regulations that are at least at stringent as the federal regulations. Second, the state must have adopted procedures to allow enforcement responsibilities to be carried out. Third, the state must keep adequate records detailing enforcement actions. If the EPA determines that the state agency has not carried out its enforcement responsibilities, EPA reports the allegation to the state. At this point the state is given 90 days to respond, after which EPA can rescind the state's enforcement authority if it is deemed necessary. State regulation of pesticides began in 1975. USEPA conducted a pilot program with six states taking over primary enforcement responsibility for FIFRA. In 1978 FIFRA was amended giving all states primary enforcement responsibility, provided their programs are approved by EPA. The enforcement responsibilities include ensuring that pesticide users follow label requirements, investigating pesticide use complaints, and inspections of pesticide users, dealers, and producers. The state agencies also have primary responsibility for training and certifying pesticide applicators. Currently, all states have enforcement responsibility and most have certification authority. The lead agency for pesticide regulation varies from state to state but it is typically the state department of agriculture. FIFRA authorizes USEPA to provide funding for state pesticide programs. Many states augment the funds with user fees such as pesticide registration fees.

Enforcement

The agreement between the EPA and the lead agency lays out a certain number of various types of investigations that must be carried out annually. These include:

- Follow up investigations: Occurs after a complaint involving pesticide application has been filed. The complaint could involve: drift of pesticides, failure to follow label directions, or human health concerns following pesticide exposure

- Pesticide use inspections: Inspections of commercial or private pesticide applicators to ensure label requirements are being followed

- Marketplace and dealer inspections: Inspections of pesticide sellers to ensure that only registered pesticides are being sold and to make sure adequate records are being kept.

- Producer establishment inspections: Inspections to ensure only registered active ingredients are being produced and that required records are being maintained.

Training and Certification

Most states have several types of commercial applicator certifications, and one type of private applicator certification. FIFRA requires that commercial applicators pass a written exam prior to receiving a license. There is no requirement that private applicators complete a written exam as part of their certification. Many agricultural states do not require private applicators to take written exams because they have many private applicators (farmers) seeking certification.

State Registration

States are allowed to register a new end use product or an additional use of a federally registered product if the situation meets the requirements specified in section 24(c) of FIFRA:

- If there is a special local need

- If the use is a food or feed use, the use must be covered by necessary tolerances or exemptions under the Federal Food, Drug and Cosmetic Act

- Registration for the use has not been previously suspended, cancelled, etc. because of health or environmental concerns

- Registration is in line with FIFRA goals

FIFRA authorizes states to issue experimental use permits (EUP), special local needs registrations (SLN) and apply for emergency exemptions. In general, states can only grant EUPs for the purpose of gathering information to support the state SLN registration process, or for experimental purposes. Section 5 of FIFRA authorizes the lead state agency to grant EUPs in accordance with the EPA-approved state plan.

Special Local Need Registrations

States are authorized by FIFRA section 24(c) to permit additional uses of federally registered pesticides through special local needs (SLN) registrations. SLN registrations are only valid in the state that issues them and must be reviewed by the EPA after the state grants the registration. The state agency must consult with EPA personnel before issuing SLN for a pesticide use that was voluntarily canceled. FIFRA section 24(c) allows states to issue SLNs under these conditions:

- There is a special local need for that product use.

- The use, if a food or feed use is covered by an appropriate tolerance or has been exempted from the requirement of a tolerance.

- The registration for the same use has not previously been denied, disapproved, suspended, or canceled by EPA, or voluntarily canceled by the registrant subsequent to EPA's issuance of a notice of intent to cancel because of health or environmental concerns about an ingredient contained in the product, unless EPA has reversed the original action.

- The registration is in accord with the purposes of FIFRA.

- If the proposed use or product falls into one of the following categories, the state must first determine that it will not cause unreasonable adverse effects on man or the environment:

  o Its composition is not similar to any federally registered product;

  o Its use pattern is not similar to any federally registered use of the same product or a product of similar composition; and

  o Other uses of the same product, or uses of a product of similar composition, have had their registration denied, disapproved, suspended, or canceled by the Administrator.

A state may register a new end-use product under one of two conditions:

- The product is identical in composition to a federally registered product but has differences in packaging, or in the identity of the formulator; or

- The product contains the same active and inert ingredients as a federally registered product, but in different percentages.

Emergency Exemptions

There are four situations in which EPA can exempt state agencies from FIFRA regulations:

1. Specific exemption

2. Quarantine exemption

3. Public health exemption

4. Crisis exemption

A specific exemption can be authorized when there is a threat of a significant economic loss, or there is significant risk to an endangered species, threatened species, beneficial organism, or the environment. The specific exemption applies for a specific time period, up to one year and can be renewed. A quarantine exemption can be authorized in order to control the spread of a pest or introduction of a pest not known to be prevalent in the concerned region. This exemption can apply for up to three years and can be renewed. A public health exemption can be authorized to control a pest believed to cause a significant risk to public health. The exemption can be authorized for up to one year, and can be renewed. A crisis exemption is allowed when the time frame between recognition of a threat and the need to act is too short to allow the state agency to obtain one of the other three exemptions. The state agency is required to inform EPA prior to issuing the emergency exemption. The duration of the exemption can be no more than 15 days, unless there is a pending specific, quarantine or public health exemption application with the EPA.

Pesticide Disposal

Pesticide disposal is handled through state programs. Most states have developed pesticide collection efforts in order to assist citizens in disposing of pesticides in an environmentally friendly way. Studies have shown that consumers store waste pesticides because they do not know the regulations for disposing of them. The Universal Waste Rule was entered into the federal register in 1995, and it provided guidelines for storage, transport, and disposal of unwanted pesticides. Many states have adopted UWR as their regulations for unwanted pesticides.

California's Regulations

California's lead agency is the California Department of Pesticide Regulation (DPR), which is within the California Environmental Protection Agency. California controls applicator licensing and pesticide registration at the state level. Enforcement and compliance of pesticide regulations occurs at the county level by the County Agricultural Commission (CAC).

In 2015, the DPR set regulation of the pesticide chloropicrin higher than the EPA. California farmers are limited to an application of chloropicrin on forty acres per day. The pesticide has caused coughing fits, irritated eyes, and headaches.

New York's Regulations

The New York State Department of Environmental Conservation is the state agency that regulates pesticides and is responsible for compliance assistance, public outreach activities and enforcement of State pesticide laws.

Department of Pesticide Regulation

- Statewide licensing programs for pesticide applicators and users.

- Evaluate and register pesticides before they can be used in California

- Perform human health risk assessments

- Perform illness surveillance

- Provide worker safety guidelines

- Test produce for pesticide residue

County Agricultural Commission

- Permit Duties

    o Evaluate permit applications to determine if proposed pesticide use will cause harm to any sensitive population or ecosystem, such as a wetland

    o Issue permits for site and time-restricted uses of restricted use pesticides

    o Require pesticide application practices to minimize risks

    o Determine whether a similar pesticide could be as effective with fewer side effects

- Ensure compliance

- Perform inspections

- Investigate pesticide injuries and illnesses

- Perform enforcement actions

    o Protect groundwater and surface water from contamination

    o Prohibit pest control company from operating in the county

    o Prohibit harvest of produce with illegal pesticide residue

    o Issue civil or criminal penalties

Restricted use pesticides are pesticides that have a higher potential for adverse impact on humans or the environment. California is the only state that requires a permit in addition to a license in order to use restricted pesticides. The county agricultural commissioner examines the permit application to determine if there is potential harm to people or the environment.

Commissioners are allowed to evaluate permits within the framework of the local conditions. Commissioners are also allowed to classify a pesticide as posing an "undue hazard" in the local environment, which requires individuals to obtain a permit in order to use that pesticide. Individuals directly impacted by the planned pesticide use can appeal the commissioner's permitting decision.

Non-regulatory Policy Mechanisms

The EPA PestWise program is a consortium of four EPA environmental stewardship programs, the Pesticide Environmental Stewardship Program, the Strategic Agriculture Initiative, the Biopesticide Demonstration Program and the Pesticide Registration Renewal Improvement Act Partnership, that work to protect human health and the environment through innovative pest management practices.

Each stewardship program focuses on accomplishing three main goals:

1. Encourage the adoption of Integrated Pest Management practices through grants and other technology transfer initiatives.

2. Provide assistance for transitioning to Integrated Pest Management practices.

3. Increase public understanding of pests and pesticide risk as well as demand for sustainable approaches to pest control.

Pesticide Environmental Stewardship Program

The Pesticide Environment Stewardship Program (PESP) was established in 1994 with the mission of reducing pesticide risk in both agricultural and non-agricultural settings through public private partnerships that promote IPM practices and the use of biological pesticides. The program is guided by the principle that the informed actions of pesticide users can potentially reduce pesticide risk more efficiently and to a greater extent than can be achieved through regulatory mandates. Additionally, PESP annually awards a maximum of $50,000 in grants to each of the ten EPA regional areas for projects that promote and support IPM practices and pesticide risk reduction activities.

Strategic Agriculture Initiative

The Strategic Agricultural Initiative (SAI) is a collaborative partnership between EPA and members of the agricultural community. SAI facilitates the transition away from the application of high-risk agricultural pesticides to using IPM methods that are cost-effective and beneficial to human health and the environment. 1.5 million dollars are distributed annually to IPM projects in the form of competitive grants. The grants are awarded through each EPA region, ensuring that the unique needs of each region are addressed appropriately. Additionally, the EPA assigns Regional Specialists that assist the agricultural growers with outreach and communication, collaborating with other stakeholders, and facilitating technology transfers. From 2003 to 2006, SAI helped agricultural growers implement IPM practices on 1,200,000 acres (4,900 km^2), leading to a 30% reduction of high-risk pesticides on those lands.

Biopesticide Demonstration Project

Established in 2003, the Biopesticides Demonstration Project (BDP) is a joint venture between the EPA and the U.S. Department of Agriculture. The overarching goal of the program is to reduce exposure to pesticides and risks of pesticide use through the increased adoption of biopesticides within the agriculture community. The BDP was developed to give agricultural growers the opportunity to observe new and innovative biopesticides across a range of agricultural conditions. To increase the awareness and share knowledge about different options for incorporating biopesticides in current farming techniques, the BDP awards competitive grants to field demonstration projects that have implemented biopesticides within an IPM system. BDP also encourages public private partnerships by providing grants to university researchers who form partnerships with agricultural growers and biopesticide companies to demonstrate the effectiveness of biopesticides. During the first five years of the program, more than 50 grants totaling 1.2 million dollars have been awarded to BDP projects. The results of these projects will be housed in a database that will be made accessible to the general public.

Pesticide Registration Renewal Improvement Act Partnership

In 2008, the Pesticide Registration Improvement Renewal Act Partnership (PRIA2) was created to promote pesticide risk reduction in the environment through the demonstration of IPM practices and technologies. PRIA2 grants a maximum of $250,000 to projects that help growers adopt IPM practices and on projects that educate our students and citizens about the benefits of IPM techniques. Since 2008, $2.3 million have been allocated to 11 grant projects. In addition to funding, PRIA2 provides assistance to the grant recipients by giving them access to data and analysis on costs associated with adopting IPM as well as measures and documents the effects of IPM programs on human health, the community and the environment.

Health and Safety

Disclosure and product safety are the difference between legal pesticide application and assault with a deadly weapon.

In most areas, physicians are mandated to file a report for "Any person suffering from any wound or other physical injury inflicted upon the person where the injury is the result of assaultive or abusive conduct." Mandated reporters are obligated to submit a report to a local law enforcement agency as follows.

1. The name and location of the injured person, if known.

2. The character and extent of the person's injuries.

3. The identity of any person the injured person alleges inflicted the wound, other injury, or assaultive or abusive conduct upon the injured person.

Employers must inform and train employees prior to pesticide exposure to avoid criminal prosecution that could occur in addition to workers compensation obligations when the employee consults a physician.

In United States common law, non-criminal battery is "harmful or offensive" contact resulting in injury that does not include intent to commit harm. This is called tortuous battery, and this falls into the same category as automobile accidents which is handled with workers compensation. This is applicable even if there is a delay between the harmful act and the resulting injury.

The definition of criminal battery is: (1) unlawful application of force (2) to the person of another (3) resulting in bodily injury. For example, a crime has been committed if the employer fails to disclose pesticide exposure in accordance with public law (unlawful force) then subsequently violates the product labeling in the assigned work area (to the person) resulting in permanent disability (bodily injury).

Pesticide injury is an accident and not a crime if the state or territorial Environment Agency is informed, employees are properly informed and trained prior to exposure, and product label restrictions are not violated. Similar principles apply to rental property occupants, occupants of public buildings like schools, customers exposed by a business owner, and the public.

Hazard Communication

A summary of workers rights is available from OSHA: Employee Workplace Rights.

The Hazard Communication Standard (HCS) first went into effect in 1985 and has since been expanded to cover almost all workplaces under OSHA jurisdiction. The details of the Hazard Communication standard are rather complicated, but the basic idea behind it is straightforward. It requires chemical manufacturers and employers to communicate information to workers about the hazards of workplace chemicals or products, including training.

The HCS was first adopted in 1983 in the United States with limited scope (48 FR 53280; November 25, 1983). In 1987, scope was expanded to cover all industries where employees are potentially exposed to hazardous chemicals (52 FR 31852; August 24, 1987). This is managed nationally within the US by the Occupational Safety and Health Administration. When a state has an approved plan, this is managed by that state instead. The standard is identified in 29 C.F.R. 1910.1200.

The United States Department of Defense manages environmental hazards in accordance with military policy that may deviate from public laws.

Employers must conduct training in a language comprehensible to employees to be in compliance with the standard. Workers must be trained at the time of initial assignment and whenever a new hazard is introduced into their work area. The purpose for this is so that workers can understand the hazards they face and so that they are aware of the protective measures that should be in place.

When OSHA conducts an inspection, the inspector will evaluate the effectiveness of the training by reviewing records of what training was done and by interviewing employees who use chemicals to find out what they understand about the hazards.

The United States Department of Transportation (DOT) regulates transportation of Dangerous Goods within the territory of the US by *Title 49 of the Code of Federal Regulations.*

All chemical manufacturers and importers must communicate hazard information through labels and Material Safety Data Sheet (MSDSs). Employers whose employees may be exposed to hazardous chemicals on the job must provide hazardous chemical information to those employees through the use of MSDSs, properly labeled containers, training, and a written hazard communication program. This standard also requires the employer to maintain a list of all hazardous chemicals used in the workplace. The MSDSs for these chemicals must be kept current and they must be made available and accessible to employees in their work areas.

Chemicals that may pose health risks or those that are physical hazards (such as fire or explosion) are covered. List of chemicals that are considered hazardous are maintained according to the use or purpose. There are several existing sources that manufacturers and employers may consult. These include:

- Any substance for which OSHA has a standard in force, including any substance listed in the Air Contaminants regulation.

- Substances listed as carcinogens (causing cancer) by the National Toxicology Program (NTP) or the International Agency for Research on Cancer (IARC).

- Substances listed in the Threshold Limit Values for Chemical Substances and Physical Agents, published by the American Conference of Governmental Industrial Hygienists (ACGIH).

- Restricted Use Products (RUP) Report; EPA

There are other sources of information about chemicals used in industry as a result of state and federal laws regarding the Community Right to Know Act.

The Air Resources Board is responsible for public hazard disclosures in California. Pesticide use disclosures are made by each pest control supervisor to the County Agricultural Commission. Epidemiology information is available from the California Pesticide Information Portal, which can be used by health care professionals to identify the cause for environmental illness.

Under the Oregon Community Right to Know Act (ORS 453.307-372) and the federal Superfund Amendments and Reauthorization Act (SARA) Title III, the Office of the State Fire Marshal collects information on hazardous substances and makes it available to emergency responders and to the general public. Among the information which companies must report are:

- Inventories of amounts and types of hazardous substances stored in their facilities.

- Annual inventories of toxic chemicals released during normal operations.

- Emergency notification of accidental releases of certain chemicals listed by the Environmental Protection Agency.

The information can be obtained in the form of an annual report of releases for the state or for specific companies. It is available on request from the Fire Marshal's Office and is normally free of charge unless unusually large quantities of data are involved.

Worker Protection Standard

The United States Environmental Protection Agency registers pesticides and insecticides as either *unclassified* or *restricted use pesticide* (RUP). Unclassified pesticide is available over-the-counter.

This licensing program exists in cooperation with the United States Department of Agriculture and state regulatory agencies.

Two items are required by law for all registered pesticides and insecticides.

- Product Label

- Material Safety Data Sheet

The product label describes how to use the product.

The Material Safety Data Sheet provides specific safety and hazard information, and this must be provided to physicians in the event that the product has been mis-used so that appropriate diagnostic testing and treatment can be obtained in a timely manner.

RUP requires license for purchase. The process required to obtain a pest control licenses is regulated by a combination of state laws, federal laws, common law, and private company policies.

All RUP applications must be recorded to identify the date, location, and type of pesticide applied. Federal law requires a minimum record retention period, which require 24 months of records to be maintained except when extended to a longer period by state laws.

There are two categories of RUP user in most areas: supervisor and applicator. A pest control supervisor license is required to purchase RUP and keep records. The pest control supervisor must ensure pest control applicators are competent to use any restricted use products. These requirements vary according to state and local law, where California has the most restrictive laws.

Further information can be obtained from the following organizations.

- AAPCO—Assoc. of American Pesticide Control Officials

- AAPSE—American Assoc. of Pesticide Safety Educators

- CTAG—Certification and Training Assessment Group

- CPARD—Certification Plan & Reporting Database

- POINTS—Pesticide of Interest Reporting Database

- NASDA Pesticide Safety Programs

- Division of Toxicology and Environmental Medicine; Agency for Toxic Substances and Disease Registry

- National Toxicology Program

40 CFR Part 170 requires the following in the United States.

- Pesticide safety training

- Notification of pesticide applications

- Use of personal protective equipment

- Restricted entry intervals following pesticide application
- Decontamination supplies
- Emergency medical assistance

Community Right to Know

Environmental health and safety outside the workplace is established by the Emergency Planning and Community Right-to-Know Act (EPCRA), which is managed by the EPA and various state and local government agencies.

State and local agencies maintain epidemiology information required by physicians to evaluate environmental illness.

Air quality information must be provided by pest control supervisors under license requirements established by the Worker Protection Standard when Restricted use pesticide is applied.

The list of restricted use pesticides is maintained by the US EPA.

Additionally, specific environmental pollutants are identified in public law, which extends to all hazardous substances even if the item is not identified as a restricted use pesticide by the EPA. As an example, cyfluthrin, cypermethrin, and cynoff contain cyanide, which is one of the most toxic known substances, but some of the products that contain these chemicals may not be identified as restricted use pesticide.

Some specific chemicals, such as cyaniate, cyanide, cyano, and nitrile compounds, satisfy the specific hazard definition that is identified in public law regardless of whether or not the item is identified on the list of restricted use pesticides maintained by the United States Environmental Protection Agency.

Most developed countries have similar regulatory practices. Pesticides and insecticides interests in the European Union are managed by the European Environment Agency.

Environmental illness share characteristics with common diseases. For example, cyanide exposure symptoms include weakness, headache, nausea, confusion, dizziness, seizures, cardiac arrest, and unconsciousness. Chest pain caused by pesticide induced bronchitis, chest pain due to asthma, and chest pain due to heart disease are all chest pain, so physicians may be unable to link pesticide exposure to environmental illness unless that information is readily available.

California requires all application by licensed pest control personnel to be disclosed, in addition to Restricted Use Pesticide. This information can be used by physicians. This is the most extensive environmental public health program in the US.

Failure of physicians to obtain pesticide disclosure will result in improper, ineffective, or delayed medical diagnosis and treatment for environmental illness caused by exposure to hazardous substance and by exposure to radiation.

U.S. Department of Transportation

The Pipeline and Hazardous Material Safety Administration within the Department of Transportation is responsible for maintaining the list of hazardous materials within the United States.

All hazardous materials that are not created at the work site must be transported by motor vehicle. The safety and security of the public transportation system is enforced by Department of Transportation.

United States Department of Transportation regulates mandatory labeling requirements for all hazardous materials. This is in addition to requirements by other federal agencies, like the United States Environmental Protection Agency, and Occupational Safety and Health Administration.

DOT is responsible for enforcement actions and public notification regarding hazardous chemical releases and exposures, including incidents involving federal workers.

DOT requires that all buildings and vehicles containing hazardous materials must have signs that disclose specific types of hazards. Hazardous materials shipper must meet the DOT's hazardous materials regulations (HMR), which are required for certified first responder to recognize unsafe conditions during hazardous substance release that can occur during transportation accidents, building fires, and weather (tornado, hurricane, ...).

Pesticide Degradation

Pesticide degradation is the process by which a pesticide is transformed into a benign substance that is environmentally compatible with the site to which it was applied. Globally, an estimated 1 to 2.5 million tons of active pesticide ingredients are used each year, mainly in agriculture. Forty percent are herbicides, followed by insecticides and fungicides. Since their initial development in the 1940s, multiple chemical pesticides with different uses and modes of action have been employed. Pesticides are applied over large areas in agriculture and urban settings. Pesticide use therefore represents an important source of diffuse chemical environmental inputs.

Persistence

In principle, pesticides are registered for use only after they are demonstrated not to persist in the environment considerably beyond their intended period of use. Typically, documented soil half-lives are in the range of days to weeks. However, pesticide residues are found ubiquitously in the environment in ng/liter to low µg/liter concentrations. For instance, surveys of groundwater and not-yet-treated potable water in industrialized countries typically detect 10 to 20 substances in recurrent findings above $0.01\ \mu g/dL$ (3.6×10^{-12} lb/cu in) the maximum accepted drinking water concentration for pesticides in many countries. About half of the detected substances are no longer in use and another 10 to 20% are stable transformation products.

Pesticide residues have been found in other realms. Transport from groundwater may lead to low-level presence in surface waters. Pesticides have been detected in high-altitude regions, demonstrating sufficient persistence to survive transport across hundreds of kilometers in the atmosphere.

Degradation involves both biotic and abiotic transformation processes. Biotic transformation is mediated by microorganisms, while abiotic transformation involves processes such as chemical and photochemical reactions. The specific degradation processes for a given pesticide are determined by its structure and by the environmental conditions it experiences. Redox gradients in soils, sediments or aquifers often determine which transformations can occur. Similarly, photochemical transformations require sunlight, available only in the topmost meter(s) of lakes or rivers, plant surfaces or submillimeter soil layers. Atmospheric phototransformation is another potential remediating influence.

Information on pesticide degradation is available from required test data. This includes laboratory tests on aqueous hydrolysis, photolysis in water and air, biodegradability in soils and water-sediment systems under aerobic and anaerobic conditions and fate in soil lysimeters. These studies provide little insight into how individual transformation processes contribute to observed degradation in situ. Therefore, they do not offer a rigorous understanding of how specific environmental conditions (e.g., the presence of certain reactants) affect degradation. Such studies further fail to cover unusual environmental conditions such as strongly sulfidic environments such as estuaries or prairie potholes, nor do they reveal transformations at low residual concentrations at which biodegradation may stop. Thus, although molecular structure generally predicts intrinsic reactivity, quantitative predictions are limited.

Biotic Transformation

Biodegradation is generally recognized as biggest contributor to degradation. Whereas plants, animals and fungi (*Eukaryota*) typically transform pesticides for detoxification through metabolism by broad-spectrum enzymes, bacteria (*Prokaryota*) more commonly metabolize them. This dichotomy is likely due to a wider range of sensitive targets in Eukaryota. For example, organophosphate esters that interfere with nerve signal transmission in insects do not affect microbial processes and offering nourishment for microorganisms whose enzymes can hydrolyze phosphotriesters. Bacteria are more likely to contain such enzymes because of their strong selection for new enzymes and metabolic pathways that supply essential nutrients. In addition, genes move horizontally within microbial populations, spreading newly evolved degradation pathways.

Some transformations, particularly substitutions, can proceed both biotically and abiotically, although enzyme-catalyzed reactions typically reach higher rates. For example, the hydrolytic dechlorination of atrazine to hydroxyatrazine in soil by atrazine-dechlorinating bacterial enzymes reached a second-order rate constant of 105/mole/second, likely dominating in the environment. In other cases, enzymes facilitate reactions with no abiotic counterpart, as with the herbicide glyphosate, which contains a C-P bond that is stable with respect to light, reflux in strong acid or base, and other abiotic conditions. Microbes that cleave the C-P bond are widespread in the environment, and some can metabolize glyphosate. The *C-P lyase* enzyme system is encoded by a complicated 14-gene operon.

Biodegradation transformation intermediates may accumulate when the enzymes that produce the intermediate operate more slowly than those that consume it. In atrazine metabolism, for example, a substantial steady-state level of hydroxyatrazine accumulates from such a process. In other situations (e.g., in agricultural wastewater treatment), microorganisms mostly grow on other, more readily assimilable carbon substrates, whereas pesticides present at trace concentrations are transformed through fortuitous metabolism, producing potentially recalcitrant intermediates.

Pesticides persist over decades in groundwater, although bacteria are in principle abundant and potentially able to degrade them for unknown reasons. This may be related to the observation that microbial degradation appears to stall at low pesticide concentrations in low-nutrient environments such as groundwater. As yet, very little is known about pesticide biodegradation under such conditions. Methods have been lacking to follow biodegradation in groundwater over the relevant long time scales and to isolate relevant degraders from such environments.

Abiotic Transformation

In surface waters, phototransformation can substantially contribute to degradation. In "direct" phototransformation, photons are absorbed by the contaminant, while in "indirect" phototransformation, reactive species are formed through photon absorption by other substances. Pesticide electronic absorption spectra typically show little overlap with sunlight, such that only a few (e.g., trifluralin) are affected by direct phototransformation. Various photochemically active light absorbers are present in surface waters, enhancing indirect phototransformation. The most prominent is dissolved organic matter (DOM), which is the precursor of excited triplet states, molecular oxygen, superoxide radical anions, and other radicals. Nitrate and nitrite ions produce hydroxyl radicals under irradiation. Indirect phototransformation is thus the result of parallel reactions with all available reactive species. The transformation rate depends on the concentrations of all relevant reactive species, together with their corresponding second-order rate constants for a given pesticide. These constants are known for hydroxyl radical and molecular oxygen. In the absence of such rate constants, quantitative structure–activity relationships(QSARs) may allow their estimation for a specific pesticide from its chemical structure.

The relevance of "dark" (aphotic) abiotic transformations varies by pesticide. The presence of functional groups supports textbook predictions for some compounds. For example, aqueous abiotic hydrolysis degrades organophosphates, carboxylic acid esters, carbamates, carbonates, some halides (methyl bromide, propargyl) and many more. Other pesticides are less amenable. Conditions such as high pH or low-redox environments combined with in situ catalyst formation including (poly)sulfides, surface-bound $Fe(II)$ or MnO_2. Microorganisms often mediate the latter, blurring the boundary between abiotic and biotic transformations. Chemical reactions may also prevail in compartments such as groundwater or lake hypolimnions, which have hydraulic retention times on the order of years and where biomass densities are lower due to almost complete absence of assimilable organic carbon.

Prediction

Available strategies to identify in situ pesticide transformation include measuring remnant or transformation product concentrations and estimation of a given environment's theoretical transformation potential. Measurements are only usable on the micro- or mesocosm scale.

Gas chromatography–mass spectrometry (GC-MS) or liquid chromatography–tandem mass spectrometry (LC-MS/MS) does not distinguish transformation from other processes such as dilution or sorption unless combined with stringent mass balance modeling. Carbon 14-labeled pesticides do enable mass balances, but investigations with radioactively tagged substrates cannot be conducted in the field.

Transformation product detection may calibrate degradation. Target analysis is straightforward when products and standards are understood, while suspect/nontarget analysis can be attempted otherwise. High-resolution mass spectrometry facilitated the development of multicomponent analytical methods for 150 pesticide transformation products and for screening for suspected transformation products. In combination with transformation product structure models, screening allows a more comprehensive assessment of transformation products, independent of field degradation studies.

Isotopic analysis may complement product measurements because it can measure degradation in the absence of metabolites and has the potential to cover sufficiently long time scales to assess transformation in groundwater. Isotope ratios (e.g.,13C/12 C, 15 N/14 N) can reveal history in the absence of any label. Because kinetic isotope effects typically favor transformation of light isotopes (e.g., 12 C), heavy isotopes (13C) become enriched in residues. An increased 13 C/12 C isotope ratio in a parent compound thus provides direct evidence of degradation. Repeated pesticide analyses, in groundwater over time, or direct measurements in combination with groundwater dating that show increasing 13 C/12 C isotope ratios in a parent pesticide, provides direct evidence of degradation, even if the pesticide was released long before. Multiple transformation pathways were revealed for atrazine by measuring isotope effects of multiple elements. In such a case, transformation mechanisms are identifiable from plots of 13 C/12 C versus 15 N/14 N parent compound data, reflecting different underlying carbon- and nitrogen-isotope effects. The approach requires a relatively high amount of substance for gas chromatography–isotope ratio mass spectrometry (GC-IRMS) or LC-IRMS analysis (100 ng to 1 μg), which, for instance, requires extraction of 10 liters of groundwater at pesticide concentrations of 100 ng/liter. For the special case of chiral pesticides, enantiomer analysis may substitute for isotopes in such analyses as a result of stereoselective reactions. Combining isotope and chirality measurement can increase prediction strength.

Geochemical analysis including pH, redox potential and dissolved ions is routinely applied to assess the potential for biotic and abiotic transformations, complicated by any lack of specificity in the targets. Selective probe compounds must be used to detect individual reactive species when a mixture of reactive species is present. Combining probe compounds and scavengers or quenchers increases accuracy. E.g., N,N-dimethylaniline, used as a probe for the carbonate radical reacts very quickly with DOM-excited triplet states and its oxidation is hampered by DOM.

13C-labeled parent pesticides were used in nontarget analysis of degraders by stable isotope probing (SIP) to demonstrate biotransformation potential in soil and sediment samples. A complementary, potentially more quantitative technique is to directly enumerate the biodegradative gene(s) via quantitative polymerase chain reaction (QPCR), gene sequencing or functional gene microarrays. A prerequisite for genetic approaches, however, is that the involved genes can be clearly linked to a given transformation reaction. For instance, the *atzD* gene encoding cyanuric acid hydrolase correlates with atrazine biodegradation in agricultural soil surface layers, consistent with *AtzD*'s cleavage of the s-triazine ring during bacterial atrazine metabolism. *AtzD* was unambiguously identifiable and hence quantifiable, as unusually, it belongs to a protein family that largely consists of biodegradative enzymes. Most proteins studied to date are members of very large protein superfamilies, with as many as 600,000 individual members, with diverse functions. Another factor confounding gene-based approaches is that biodegradative function can arise independently in evolution, such that multiple unrelated genes catalyze the same reaction. E.g., organophosphate esterases that differ markedly in their fold and mechanism can act on the same organophosphate pesticide.

Transformation Products

Even though their undesirable effects are typically lowered, transformation products may remain problematic. Some transformations leave active moiety intact, such as oxidation of thioethers to sulfones and sulfoxides. Parent/transformation product mixtures may have additive effects. Second, some products are more potent than their parents. Phenolic degradates of such diverse chemical classes as pyrethroids and aryloxyphenoxypropionic herbicides may act on estrogen receptor. Such products should receive particular attention because they are often smaller and more polar than their parents. This increases their potential to reach drinking water resources such as groundwater and surface waters, where polar products are found at fairly constant concentrations. Products in drinking water resources may cause problems such as formation of carcinogenic N-nitroso-dimethylamine from dimethylsulfamide, a microbial product of the fungicides tolylfluanide and dichlofluanide, during water treatment with ozone.

The issue is specifically addressed in major regulatory frameworks. In Europe, for instance, "nonrelevant" metabolites are distinguished from metabolites that are "relevant for groundwater resources" or even "ecotoxicologically relevant". The latter are those whose risk to soil or aquatic biota is comparable or higher than the parent and must meet the same standards as their parent. Groundwater-relevant metabolites are those likely to reach groundwater in concentrations above 0.1 µg/liter and to display the same toxicity as the parent compound. In the past toxicology issues typically emerged only decades after market introduction. Examples are the detection of chloridazon products (first marketed in 1964) in surface and groundwater, or tolylfluanid (first marketed in 1971). That these substances were overlooked for so long may partially be attributable to prior limits on analytical capabilities. However, labeling some metabolites as nonrelevant may have resulted in directing attention away from them. The decision to tolerate up to 10 µg/liter of "nonrelevant" metabolites in groundwater and drinking water is politically highly contentious in Europe. Some consider the higher limit acceptable as no imminent health risk can be proven, whereas others regard it as a fundamental deviation from the precautionary principle.

References

- Marten, Gerry "Non-pesticide management" for escaping the pesticide trap in Andrah Padesh, India. ecotippingpoints.org. Retrieved on September 17, 2007.

- Marino M. (August 2007), Blowies inspire pesticide attack: Blowfly maggots and dog-wash play starring roles in the story of a remarkable environmental clean-up technology. Solve, Issue 12. CSIRO Enquiries. Retrieved on 2007-10-03.

- Graeme Murphy (December 1, 2005), Resistance Management - Pesticide Rotation. Ontario Ministry of Agriculture, Food and Rural Affairs. Retrieved on September 15, 2007

- Stephen J. Toth, Jr., Pesticide Impact Assessment Specialist, North Carolina Cooperative Extension Service, "Federal Pesticide Laws and Regulations." March, 1996. Retrieved on February 25, 2011.

- Smith, Scott (14 January 2015). "California Unveils Strictest Rules In U.S. For Common Pesticide". The Huffington Post. Retrieved 15 January 2015.

- "New York State Department of Environmental Conservation". Herkimer Evening Telegram. 13 January 2014. Retrieved 13 January 2014.

- Hoigné, J. (1990). Werner Stumm, ed. Reaction Rates of Processes in Natural Waters. Aquatic chemical kinetics: reaction rates of processes in natural waters. Wiley. ISBN 978-0-471-51029-1.

Pesticide Applications: An Overview

To ensure that pesticides are delivered to the target organisms efficiently and to minimize the potential risk to people and the environment, the method of pesticide application is as important as the right pesticide. This section explores topics such as aerosol sprays, aerial application and ultra-low volume pesticide application. These techniques improve targeting and decrease spraying inefficiencies.

Pesticide Application

Pesticide application refers to the practical way in which pesticides, (including herbicides, fungicides, insecticides, or nematode control agents) are delivered to their *biological targets* (*e.g.* pest organism, crop or other plant). Public concern about the use of pesticides has highlighted the need to make this process as efficient as possible, in order to minimize their release into the environment and human exposure (including operators, bystanders and consumers of produce). The practice of pest management by the rational application of pesticides is supremely multi-disciplinary, combining many aspects of biology and chemistry with: agronomy, engineering, meteorology, socio-economics and public health, together with newer disciplines such as biotechnology and information science.

Seed Treatments

Seed treatments can achieve exceptionally high efficiencies, in terms of effective dose-transfer to a crop. Pesticides are applied to the seed prior to planting, in the form of a seed treatment, or coating, to protect against soil-borne risks to the plant; additionally, these coatings can provide supplemental chemicals and nutrients designed to encourage growth. A typical seed coating can include a nutrient layer—containing nitrogen, phosphorus, and potassium, a rhizobial layer—containing symbiotic bacteria and other beneficial microorganisms, and a fungicide (or other chemical) layer to make the seed less vulnerable to pests.

Spray Application

One of the most common forms of pesticide application, especially in conventional agriculture, is the use of mechanical sprayers. Hydraulic sprayers consists of a tank, a pump, a lance (for single nozzles) or boom, and a nozzle (or multiple nozzles). Sprayers convert a pesticide formulation, often containing a mixture of water (or another liquid chemical carrier, such as fertilizer) and chemical, into droplets, which can be large rain-type drops or tiny almost-invisible particles. This conversion is accomplished by forcing the spray mixture through a spray nozzle under pressure. The size of droplets can be altered through the use of different nozzle sizes, or by altering the pressure under which it is forced, or a combination of both. Large droplets have the advantage of being less susceptible to spray drift, but require more water per unit of land covered. Due to static electricity, small droplets are able to maximize contact with a target organism, but very still wind conditions are required.

Spraying Pre- and Post-Emergent Crops

Traditional agricultural crop pesticides can either be applied pre-emergent or post-emergent, a term referring to the germination status of the plant. Pre-emergent pesticide application, in conventional agriculture, attempts to reduce competitive pressure on newly germinated plants by removing undesirable organisms and maximizing the amount of water, soil nutrients, and sunlight available for the crop. An example of pre-emergent pesticide application is atrazine application for corn. Similarly, glyphosate mixtures are often applied pre-emergent on agricultural fields to remove early-germinating weeds and prepare for subsequent crops. Pre-emergent application equipment often has large, wide tires designed to float on soft soil, minimizing both soil compaction and damage to planted (but not yet emerged) crops. A three-wheel application machine, such as the one pictured on the right, is designed so that tires do not follow the same path, minimizing the creation of ruts in the field and limiting sub-soil damage.

Large self-propelled agricultural 'floater' sprayer, engaged in pre-emergent pesticide application

Post-emergent pesticide application requires the use of specific chemicals chosen minimize harm to the desirable target organism. An example is 2,4-Dichlorophenoxyacetic acid, which will injure broadleaf weeds (dicots) but leave behind grasses (monocots). Such a chemical has been used extensively on wheat crops, for example. A number of companies have also created genetically modified organisms that are resistant to various pesticides. Examples include glyphosate-resistant soybeans and Bt maize, which change the types of formulations involved in addressing post-emergent pesticide pressure. It is important to also note that even given appropriate chemical choices, high ambient temperatures or other environmental influences, can allow the non-targeted desirable organism to be damaged during application. As plants have already germinated, post-emergent pesticide application necessitates limited field contact in order to minimize losses due to crop and soil damage. Typical industrial application equipment will utilize very tall and narrow tires and combine this with a sprayer body which can be raised and lowered depending on crop height. These sprayers usually carry the label 'high-clearance' as they can rise over growing crops, although usually not much more than 1 or 2 meters high. In addition, these sprayers often have very wide booms in order to minimize the number of passes required over a field, again designed to limit crop damage and maximize efficiency. In industrial agriculture, spray booms 120 feet (40 meters) wide are not uncommon, especially in prairie agriculture with large, flat fields. Related to this, aerial pesticide application is a method of top dressing a pesticide to an emerged crop which eliminates physical contact with soil and crops.

Self-propelled row-crop sprayer applying pesticide to post-emergent corn

Air Blast sprayers, also known as air-assisted or mist sprayers, are often used for tall crops, such as tree fruit, where boom sprayers and aerial application would be ineffective. These types of sprayers can only be used where overspray—spray drift—is less of a concern, either through the choice of chemical which does not have undesirable effects on other desirable organisms, or by adequate buffer distance. These can be used for insects, weeds, and other pests to crops, humans, and animals. Air blast sprayers inject liquid into a fast-moving stream of air, breaking down large droplets into smaller particles by introducing a small amount of liquid into a fast-moving stream of air.

Foggers fulfill a similar role to mist sprayers in producing particles of very small size, but use a different method. Whereas mist sprayers create a high-speed stream of air which can travel significant distances, foggers use a piston or bellows to create a stagnant area of pesticide that is often used for enclosed areas, such as houses and animal shelters.

Spraying Inefficiencies

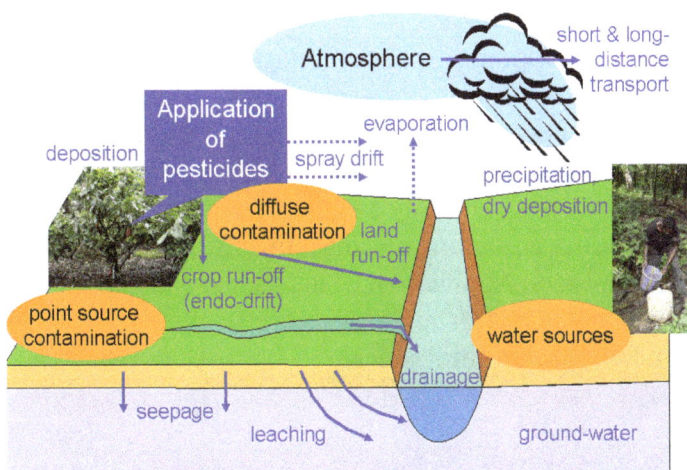

In order to better understand the cause of the spray inefficiency, it is useful to reflect on the implications of the large range of droplet sizes produced by typical (hydraulic) spray nozzles. This has long been recognized to be one of the most important concepts in spray application (*e.g.* Himel, 1969), bringing about enormous variations in the properties of droplets.

Historically, dose-transfer to the biological target (*i.e.* the pest) has been shown to be inefficient. However, relating "ideal" deposits with biological effect is fraught with difficulty), but in spite of Hislop's misgivings about detail, there have been several demonstrations that massive amounts of pesticides are wasted by run-off from the crop and into the soil, in a process called endo-drift. This is a less familiar form of pesticide drift, with exo-drift causing much greater public concern. Pesticides are conventionally applied using hydraulic atomisers, either on hand-held sprayers or tractor booms, where formulations are mixed into high volumes of water.

Different droplet sizes have dramatically different dispersal characteristics, and are subject to complex macro- and micro-climatic interactions (Bache & Johnstone, 1992). Greatly simplifying these interactions in terms of droplet size and wind speed, Craymer & Boyle concluded that there are essentially three sets of conditions under which droplets move from the nozzle to the target. These are where:

- sedimentation dominates: typically larger (>100 µm) droplets applied at low wind-speeds; droplets above this size are appropriate for minimising drift contamination by herbicides.

- turbulent eddies dominate: typically small droplets (<50 µm) that are usually considered most appropriate for targeting flying insects, unless an electrostatic charge is also present that provides the necessary force to attract droplets to foliage. (NB: the latter effects only operate at very short distances, typically under 10 mm.)

- intermediate conditions where both sedimentation and drift effects are important. Most agricultural insecticide and fungicide spraying is optimised by using relatively small (say 50-150 µm) droplets in order to maximize "coverage" (droplets per unit area), but are also subject to drift.

Herbicide Volatilisation

Herbicide volatilisation refers to evaporation or sublimation of a volatile herbicide. The effect of gaseous chemical is lost at its intended place of application and may move downwind and affect other plants not intended to be affected causing crop damage. Herbicides vary in their susceptibility to volatilisation. Prompt incorporation of the herbicide into the soil may reduce or prevent volatilisation. Wind, temperature, and humidity also affect the rate of volatilisation with humidity reducing in. 2,4-D and dicamba are commonly used chemicals that are known to be subject to volatilisation but there are many others. Application of herbicides later in the season to protect herbicide-resistant genetically modified plants increases the risk of volatilisation as the temperature is higher and incorporation into the soil impractical.

Improved Targeting

In the 1970s and 1980s improved application technologies such as controlled droplet application (CDA) received extensive research interest, but commercial uptake has been disappointing. By controlling droplet size, ultra-low volume (ULV) or very low volume (VLV) application rates of pesticidal mixtures can achieve similar (or sometimes better) biological results by improved timing and dose-transfer to the biological target (*i.e.* pest). No atomizer has been developed able to produce uniform (monodisperse) droplets, but rotary (spinning disc and cage) atomizers usually produce a more uniform droplet size spectrum than conventional hydraulic nozzles. Other efficient application techniques include: banding, baiting, specific granule placement, seed treatments and weed wiping.

The Ulvamast Mk II: a ULV sprayer for locust control (photo taken in Niger)

CDA is good examples of a rational pesticide use (RPU) technology (Bateman, 2003), but unfortunately has been unfashionable with public funding bodies since the early 1990s, with many believing that all pesticide development should be the responsibility of pesticide manufacturers. On the other hand, pesticide companies are unlikely widely to promote better targeting and thus reduced pesticide sales, unless they can benefit by adding value to products in some other way. RPU contrasts dramatically with the promotion of pesticides, and many agrochemical concerns, have equally become aware that product stewardship provides better long-term profitability than high pressure salesmanship of a dwindling number of new "silver bullet" molecules. RPU may therefore provide an appropriate framework for collaboration between many of the stake-holders in crop protection.

Understanding the biology and life cycle of the pest is also an important factor in determining droplet size. The Agricultural Research Service, for example, has conducted tests to determine the ideal droplet size of a pesticide used to combat corn earworms. They found that in order to be effective, the pesticide needs to penetrate through the corn's silk, where the earworm's larvae hatch. The research concluded that larger pesticide droplets best penetrated the targeted corn silk. Knowing where the pest's destruction originates is crucial in targeting the amount of pesticide needed.

Quality and Assessment of Equipment

Ensuring quality of sprayers by testing and setting of standards for application equipment is important to ensure users get value for money. Since most equipment uses various hydraulic nozzles, various initiatives have attempted to classify spray quality, starting with the BCPC system.

IPARC houses and carries out the World Health Organisation fatigue test for pressurised equipment: used for indoor residue spraying (IRS) against mosquitoes, other disease vectors and (sometimes) in agriculture

Other Application Methods

- Granule application equipment

- Dust application

Application Methods for Household Insecticides

Pest management in the home begins with restricting the availability to insects of three vital commodities: shelter, water and food. If insects become a problem despite such measures, it may become necessary to control them using chemical methods, targeting the active ingredient to the particular pest. Insect repellent, referred to as "bug spray", comes in a plastic bottle or aerosol can. Applied to clothing, arms, legs, and other extremities, the use of these products will tend to ward off nearby insects. This is not an insecticide.

Insecticide used for killing pests—most often insects, and arachnids—primarily comes in an aerosol can, and is sprayed directly on the insect or its nest as a means of killing it. Fly sprays will kill house flies, blowflies, ants, cockroaches and other insects and also spiders. Other preparations are granules or liquids that are formulated with bait that is eaten by insects. For many household pests bait traps are available that contain the pesticide and either pheromone or food baits. Crack and crevice sprays are applied into and around openings in houses such as baseboards and plumbing. Pesticides to control termites are often injected into and around the foundations of homes.

Active ingredients of many household insecticides include permethrin and tetramethrin, which act on the nervous system of insects and arachnids.

Bug sprays should be used in well ventilated areas only, as the chemicals contained in the aerosol and most insecticides can be harmful or deadly to humans and pets. All insecticide products including solids, baits and bait traps should be applied such that they are out of reach of wildlife, pets and children.

Aerial Application |

Aerial application, or what was formerly referred to as crop dusting, involves spraying crops with crop protection products from an agricultural aircraft. Planting certain types of seed are also included in aerial application. The specific spreading of fertilizer is also known as *aerial topdressing* in some countries.

Agricultural aircraft are highly specialized, purpose-built aircraft. Today's agricultural aircraft are often powered by turbine engines of up to 1500 hp and can carry as much as 800 gallons of crop protection product. Helicopters are sometimes used, and some aircraft serve double duty as water bombers in areas prone to wildfires.(These aircraft are referred to as S.E.A.T. "single engine air tankers").

History

Aerial Seed Sowing 1906

The first known aerial application of agricultural materials was by John Chaytor, who in 1906 spread seed over a swamped valley floor in Wairoa, New Zealand, using a hot air balloon with mobile tethers. Aerial sowing of seed still continues to this day with cover crop applications and rice planting.

Crop Dusting 1921

The first known use of a heavier-than-air machine to disperse products occurred on 3 August 1921. Crop dusting was developed under the joint efforts of the U.S. Agriculture Department, and the U.S. Army Signal Corps's research station at McCook Field in Dayton, Ohio. Under the direction of McCook engineer Etienne Dormoy, a United States Army Air Service Curtiss JN4 Jenny piloted by John A. Macready was modified at McCook Field to spread lead arsenate to kill catalpa sphinx caterpillars at a Catalapa farm near Troy, Ohio in the United States. The first test was considered highly successful. The first commercial operations were begun in 1924, by Huff-Daland Crop Dusting, which was co-founded by McCook Field test pilot Lt. Harold R. Harris. Use of insecticide and fungicide for crop dusting slowly spread in the Americas and to a lesser extent other nations in the 1930s. The name 'crop dusting' originated here, as actual dust was spread across the crops. Today, aerial applicators use liquid crop protection products in very small doses.

Lt. Macready (right) and McCook Field engineer E. Dormoy (left) in front of the 1st crop duster airplane (August 3, 1921)

Top Dressing 1939–1946

Aerial topdressing, the spread of fertilizers such as superphosphate, was developed in New Zealand in the 1940s by members of the Ministry of Public Works and RNZAF, led by Alan Pritchard and Doug Campbell - unofficial experiments by individuals within the government led to funded research. Initially fertilizer and seed were dropped together (1939), using a window mounted chute on a Miles Whitney Straight, but by the end of the 1940s different mixtures of fertilizer were being distributed from hoppers installed in war surplus Grumman Avengers and C-47 Skytrains, as well as some privately operated de Havilland Tiger Moths in New Zealand, and the practise was being adopted experimentally in Australia and the United Kingdom.

Purpose Built Aircraft

In 1951, Leland Snow begins designing the first aircraft specifically built for aerial application, the S-1.

In 1957, The Grumman G-164 Ag-Cat is the first aircraft designed by a major company for ag aviation.

Water Bombing 1952

Aerial firefighting, or water bombing, was tested experimentally by Art Seller's Skyways air services in Canada in 1952 (dropping a mix of water, fertilizer and seed), and established in California in the mid-1950s.

Night Aerial Application 1973–Present

Aerial application at night is mostly liquid spray and is conducted in the Southwest U.S. deserts. The increased heat, scheduling conflicts with farm workers in the fields and honeybee activity reduced the effectiveness of spraying in daytime. In high temperature areas, the insects would travel down in plants in daytime and return to the top at night. The aircraft — both fixed wing, autogyros and helicopters — were equipped with lights, usually three sets: Work lights were high power and aimed or adjustable from the cockpit; wire lights were angled down for taxiing and wire or obstruction illumination; and turn lights were only turned on in the direction of the turn to allow safe operation on moonless nights where angle of entry or exit needed to be illuminated. Some aircraft were equipped with an elongated metal wing called a spreader, with inbuilt channels to direct the flow of dust such as sulfur, used on melons as a pesticide and soil amendment. Very little pesticide dust was used day or night in comparison to spray, because of the difficulty in drift control. Workers on the ground, called "flaggers", would use flashlights aimed at the aircraft to mark the swaths on the ground; later, GPS units replaced the flaggers due to new laws restricting use of human flaggers with some pesticides. GPS systems also provide precise guidance for the applicator.

Agricultural chemicals have also kept pace with advancements in technology, and have been influential in the growth of the agricultural aviation industry. In the 1930s Aerial Applicators arrived in the northern states to war against insect and disease pests which threatened fruit and vegetable crops. After World War II, the industry expanded into the western states where the development of new chemicals made possible the control of weeds and insects in cereal grain crops. Some of these new chemicals proved very useful in controlling various insects that carried diseases dangerous to humans. Countries that previously had no control over malaria and river blindness were provided with chemicals which helped save hundreds of thousands of lives and reduced the suffering of millions. All during the 1950s, crop production continued to rise and disease declined as a result of chemical controls.

Aerial application accounts for just under 20% of all applied crop protection products on commercial farms. The industry also provides firefighting and public health application services According to a 2012 NAAA survey, the five most common aerially treated crops are: corn, wheat/barley, soybeans, pastures/rangelands and alfalfa, but aerial application is used on many more crops grown in the U.S.

Approximately 1,350 aerial application businesses are in the U.S. and 1,430 non-operator pilots. 94% of aerial application business owners (operators) are also pilots. Aerial application businesses are located in 44 states – all but Connecticut, Hawaii, Nevada, Rhode Island, Vermont and West Virginia.

Today's ag aircraft use sophisticated precision application equipment such as: GPS (global positioning systems), GIS (geographical information systems), Aircraft Integrated Meteorological Measurement System (AIMMS), real time meteorological systems, flow control valves for variable-rate applications, single-boom shutoff valves and smokers to identify wind speed and direction.

According to the U.S. Bureau of Labor Statistics, in 2005 U.S. cropduster pilots earned an average annual wage of $63,210.

Unmanned Aerial Application

Beginning in the late 1990s, unmanned aerial vehicles are also being used for agricultural spraying. This phenomenon started in Japan and South Korea, where mountainous terrain and relatively small family-owned farms required lower-cost and higher precision spraying. As of 2014, the use of UAV crop dusters, such as the Yamaha R-MAX, is being expanded to the United States for use in spraying of vineyards.

A Yamaha R-MAX, a UAV commonly used for aerial application in Japan.

Ultra-low Volume

The term Ultra-Low Volume (ULV) (spraying) is used in the context of pesticide application.

Ultra-low volume application of pesticides has been defined as spraying at a Volume Application Rate (VAR) of less than 5 L/ha for field crops or less than 50 L/ha for tree/bush crops. VARs of 0.25 – 2 l/ha are typical for aerial ULV application to forest or migratory pests.

ULV spraying is a well-established spraying technique and remains the standard method of locust control with pesticides and is also widely used by cotton farmers in central-southern and western Africa. It has also been used in massive aerial spraying campaigns against disease vectors such as the tse-tse fly.

A major benefit of ULV application is high work rate (i.e. hectares can be treated in one day). It is a good option if all (or some) of these conditions apply:

- large area of land to treat
- rapid response required
- little or no water for making pesticide tank mixtures
- logistical problems for supplies
- difficult terrain: poor access to target site

Equipment

ULV equipment is designed to produce very small droplets, thus ensuring even coverage with low volumes. The equipment is based on aerosol, air-shear (mistblowers, exhaust gas sprayers) or better still, rotary nozzle techniques. An electrostatic charge may be applied to the droplets to aid their distribution and impaction (on earthed targets), but commercial equipment is rare at present.

Ultra Low Volume Fogging Machines

Ultra low volume (ULV) fogging machines are cold fogging machines that use large volumes of air at low pressures to transform liquid into droplets that are dispersed into the atmosphere. This type of fogging machine can produce extremely small droplets with diameters ranging from 1-150 μm. ULV machines are used for applying pesticides, herbicides, fungicides, sterilizers, and disinfectants amongst other chemicals. The size of the droplet is very important as each application has an optimal droplet size. The optimum droplet sizes are between 5 and 20 μm for flying insects, 20 to 40 μm for leaf nematodes and 30 to 50 μm for fungi. Low volume refers to the low volume of carrier fluid that is required with these types of machines. The droplets that are created are of such a small size that less carrier for the formulation is required to cover the required surface area. The best way to understand the concept of using less formulation to cover a larger surface area is to look at the mathematical side of the scenario. In the case where the diameter of a droplet is reduced to half its original size then the amount of droplets that can be formed from the same volume of formulation will increase eightfold. If the droplet diameter is reduced to 10 percent of its original size, then the amount of droplets that can be formed will increase a thousandfold. In this way the droplet diameter determines the amount of droplets that will form.

ULV Fogging Machine Parts

Ultra low volume fogging machines consists of a blower, a formulation-holding tank and in some equipment a pump. The machine can have an electric, battery or gasoline engine that drives the blower. The blower creates a low pressure area and forces air through the nozzle of the fog machine. Air pressure can be controlled by adjusting the engine speed. Formulation is delivered by means of either electric, gear, FMI or Diaphragm pump to deliver the formulation to the nozzle of the machine, or in other equipment it is delivered through creating a low air pressure in the formulation tank to force the formulation to the nozzle for easy application. The nozzle of the machine has a very specific shape, which causes a swirling motion of the air stream. The motion is achieved by means of several stationary fins that force the air to rotate. The formulation is delivered to the air by means of a supply tube that is situated in the center of the nozzle. The motion of the air shears the liquid formulation into very small droplets and then disperses it into the atmosphere.

Typical large ULV cold fogging machine. Max Pro 145 model from Dyna-fog Africa

ULV fogging machines are the most preferred misting machines as the size of the droplets can be controlled by means of machine calibration.

ULV Fogging Machine Advantages and Disadvantages

The chemicals dispersed in this type of machine are more concentrated than the chemicals used in other spraying equipment, which also increases the killing efficiency. Other advantages of ULV misting machines includes lower risks of injury due to the fog cloud being nearly invisible, low volumes of carrier chemicals, lower application cost and low noise levels. Unfavorable aspects of these machines may include longer application times, wind drift, high concentrations of active ingredients causing environmental hazards, and the requirement of higher technical skills for calibration of the machines.

ULV Fogging Machine Applications

ULV fogging machines can be used in a number of different industries. Some applications includes pest control: mosquito control, bird control, agricultural applications such as grain storage, disinfectant purposes such as hospitals and laboratories, mold control and surface decontamination. A specific application for ULV machines that have been well researched is protecting avocado trees from different diseases. The most common diseases that these trees are prone to suffer from includes Cercospora spot, anthracnose and stem-end rot. The diseases affecting the avocado trees are controlled by applying high volume copper oxychloride fungicides to the trees. The original application techniques included the use of a hand gun sprayer. This technique posed the problem of high run-off of the formulation. The use of ULV machines for the application of the pesticide formulation yielded more than 80 percent healthy fruit that was free from Cercospora spot. These results compared very favorably with the traditional method of using a hand gun sprayer. Another industry that have benefited substantially from the technology provided by ULV fogging machines is the chicken industry. This industry suffers great losses due to the litter beetle and *Aspergillus* fungi. ULV fogging machines offers great solutions to kill both these pests in chicken houses.

Aerosol Spray

Aerosol spray is a type of dispensing system which creates an aerosol mist of liquid particles. This is used with a can or bottle that contains a payload and propellant under pressure. When the container's valve is opened, the payload is forced out of a small hole and emerges as an aerosol or mist. As propellant expands to drive out the payload, only some propellant evaporates inside the can to maintain a constant pressure. Outside the can, the droplets of propellant evaporate rapidly, leaving the payload suspended as very fine particles or droplets. Typical payload liquids dispensed in this way are insecticides, deodorants and paints.

Aerosol spray can

An atomizer is a similar device that is pressurised by a hand-operated pump rather than by stored propellant.

History

The concepts of aerosol probably go as far back as 1790. The first aerosol spray can patent was granted in Oslo in 1927 to Erik Rotheim, a Norwegian chemical engineer, and a United States patent was granted for the invention in 1931. The patent rights were sold to a United States company for 100,000 Norwegian kroner. The Norwegian Postal Service, Posten Norge, celebrated the invention by issuing a stamp in 1998.

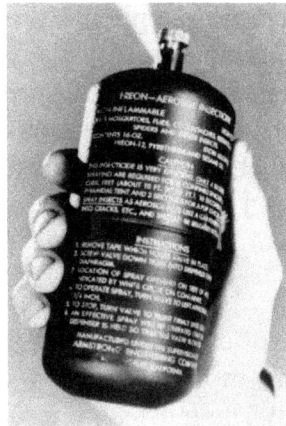

The aerosol (A gaseous suspension of fine solid or liquid particles) spray canister invented by USDA researchers, Lyle Goodhue and William Sullivan.

In 1939, American Julian S. Kahn received a patent for a disposable spray can, but the product remained largely undeveloped. Kahn's idea was to mix cream and a propellant from two sources to make whipped cream at home — not a true aerosol in that sense. Moreover, in 1949, he disclaimed his first four claims, which were the foundation of his following patent claims. It was not until 1941 that the aerosol spray can was first put to good use by Americans Lyle Goodhue and William Sullivan, who are credited as the inventors of the modern spray can. Their design of a refillable spray can dubbed the "bug bomb", is the ancestor of many popular commercial spray products. Pressurized by liquefied gas, which gave it propellant qualities, the small, portable can enabled soldiers to defend against malaria-carrying mosquitoes by spraying inside tents and airplanes in the Pacific during World War II. Goodhue and Sullivan received the first Erik Rotheim Gold Medal from the Federation of European Aerosol Associations on August 28, 1970 in Oslo, Norway in recognition of their early patents and subsequent pioneering work with aerosols. In 1948, three companies were granted licenses by the United States government to manufacture aerosols. Two of the three companies, Chase Products Company and Claire Manufacturing, still manufacture aerosols to this day. The "crimp-on valve", used to control the spray in low-pressure aerosols was developed in 1949 by Bronx machine shop proprietor Robert H. Abplanalp.

In 1974, Drs. Frank Sherwood Rowland and Mario J. Molina proposed that chlorofluorocarbons, used as propellants in aerosol sprays, contributed to the depletion of Earth's ozone layer. In response to this theory, the U.S. Congress passed amendments to the Clean Air Act in 1977 authorizing the Environmental Protection Agency to regulate the presence of CFCs in the atmosphere. The United Nations Environment Programme called for ozone layer research that same year, and, in 1981, authorized a global framework convention on ozone layer protection. In 1985, Joe Farman, Brian G. Gardiner, and Jon Shanklin published the first scientific paper detailing the hole in the ozone layer. That same year, the Vienna Convention was signed in response to the UN's authorization. Two years later, the Montreal Protocol, which regulated the production of CFCs was formally signed. It came into effect in 1989. The U.S. formally phased out CFCs in 1995.

Aerosol Propellants

If aerosol cans were simply filled with compressed gas, it would either need to be at a dangerously high pressure and require special pressure vessel design (like in gas cylinders), or the amount of payload in the can would be small, and rapidly deplete. Usually the gas is the vapor of a liquid with boiling point slightly lower than room temperature. This means that inside the pressurized can, the vapor can exist in equilibrium with its bulk liquid at a pressure that is higher than atmospheric pressure (and able to expel the payload), but not dangerously high. As gas escapes, it is immediately replaced by evaporating liquid. Since the propellant exists in liquid form in the can, it should be miscible with the payload or dissolved in the payload. In gas dusters, the propellant itself acts as the payload. The propellant in a gas duster can is not "compressed air" as sometimes assumed, but usually a haloalkane.

Chlorofluorocarbons (CFCs) were once often used as propellants, but since the Montreal Protocol came into force in 1989, they have been replaced in nearly every country due to the negative effects CFCs have on Earth's ozone layer. The most common replacements are mixtures of volatile hydrocarbons, typically propane, n-butane and isobutane. Dimethyl ether (DME) and methyl ethyl ether are also used. All these have the disadvantage of being flammable. Nitrous oxide and carbon dioxide are also used as propellants to deliver foodstuffs (for example, whipped cream and cooking spray). Medicinal aerosols such as asthma inhalers use hydrofluoroalkanes (HFA): either HFA 134a (1,1,1,2,-tetrafluoroethane) or HFA 227 (1,1,1,2,3,3,3-heptafluoropropane) or combinations of the two. Manual pump sprays can be used as an alternative to a stored propellant.

The above situation may change as new technology emerges. A new patented family of aerosol valves has been developed by the Spray Research Group at Salford University (UK), and these valves generate a 'bubbly flow' within aerosol cans, using compressed air or inert gas. This technology is now being brought to market by The Salford Valve Company ('Salvalco'). 'Packaging Today' & 'Packaging Europe News'6 March 2015

Another UK company (42 Technology) has developed a patented technology to generate more finely dispersed mists by using a disc of superhydrophobic material within the manual pump.

Packaging

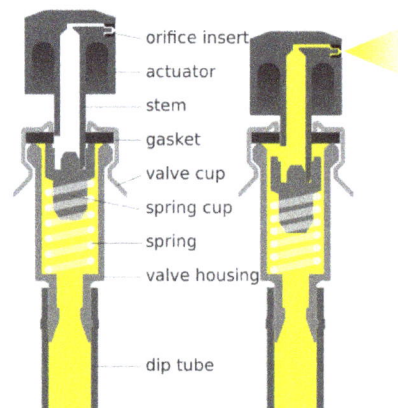

orifice insert
actuator
stem
gasket
valve cup
spring cup
spring
valve housing
dip tube

A typical paint valve system will have a "female" valve, the stem being part of the top actuator. The valve can be preassembled with the valve cup and installed on the can as one piece, prior to pressure-filling. The actuator is added later.

Modern aerosol spray products have three major parts: the can, the valve and the actuator or button. The can is most commonly lacquered tinplate (steel with a layer of tin) and may be made of two or three pieces of metal crimped together. Aluminium cans are also common and are generally used for more expensive products. The valve is crimped to the rim of the can, and the design of this component is important in determining the spray rate. The actuator is depressed by the user to open the valve; a spring closes the valve again when it is released. The shape and size of the nozzle in the actuator controls the spread of the aerosol spray.

Non-aerosol Packaging Alternatives

By definition, aerosol sprays release their propellant during use. Some non-aerosol alternatives include the following denoted below.

Packaging that uses a piston barrier system by CCL Industries or EarthSafe by Crown Holdings is often selected for highly viscous products such as post-foaming hair gels, thick creams and lotions, food spreads and industrial products and sealants. The main benefit of this system is that it eliminates gas permeation and assures separation of the product from the propellant, maintaining the purity and integrity of the formulation throughout its consumer lifespan. The piston barrier system also provides a consistent flow rate with minimal product retention.

Another type of dispensing system is the bag-in-can (or BOV, bag-on-valve technology) system where the product is separated from the pressurizing agent with a hermetically sealed, multi-layered laminated pouch, which maintains complete formulation integrity so only pure product is dispensed. Among its many benefits, the bag-in-can system extends a product's shelf life, is suitable for all-attitude, (360-degree) dispensing, quiet and non-chilling discharge. This bag-in-can system is used in the packaging of pharmaceutical, industrial, household, pet care and other products that require complete separation between the product and the propellant.

A new development is the 2K (two component) aerosol. A 2K aerosol device has main component stored in main chamber and a second component stored in an accessory container. When applicator activates the 2K aerosol by breaking the accessory container, the two components mix. The 2K aerosol can has the advantage for delivery of reactive mixtures. For example, 2K reactive mixture can use low molecular weight monomer, oligomer, and functionalized low molecular polymer to make final cross-linked high molecular weight polymer. 2K aerosol can increase solid contents and deliver high performance polymer products, such as curable paints, foams, and adhesives.

Safety Concerns

Canned air / dusters do *not* contain air, and are dangerous, even deadly, to inhale.

There are three main areas of health concern linked to aerosol cans:

- Aerosol contents may be deliberately inhaled to achieve intoxication from the propellant (known as inhalant abuse or "huffing"). Calling them "canned air" or "cans of compressed air" could mislead the ignorant to think they are harmless. Snopes has multiple reports of deaths from misuse.

- Aerosol burn injuries can be caused by the spraying of aerosol directly onto the skin, in a practice sometimes called "frosting". Adiabatic expansion causes the aerosol contents to cool rapidly on exiting the can.

- The propellants in aerosol cans are typically combinations of ignitable gases and have been known to cause fires and explosions.

In the United States, non-empty aerosol cans are considered hazardous waste.

References

- Waxman, Michael F., (1998) Application Equipment. In: Agrochemical and Pesticide Safety Handbook Ed. M. Wilson. CRC Press, Boca Raton (ISBN 978-1-56670-296-6).

- Andrew Pollack (April 25, 2012). "Dow Corn, Resistant to a Weed Killer, Runs Into Opposition". The New York Times. Retrieved April 25, 2012.

- Fabian Menalled and William E. Dyer. "Getting the Most from Soil-Applied Herbicides". Montana State University. Retrieved April 25, 2012.

- "Before spraying wildly at anything that moves, consider more reasoned approach". reviewjournal.com. Retrieved 23 February 2014.

- Kvilesjø, Svend Ole (17 February 2003). "Sprayboksens far er norsk". Aftenposten (in Norwegian). Archived from the original on 30 June 2008. Retrieved 6 February 2009.

- Carlisle, Rodney (2004). Scientific American Inventions and Discoveries, p.402. John Wiley & Songs, Inc., New Jersey. ISBN 0-471-24410-4.

- Kimberley A. McGrath (Editor), Bridget E. Travers (Editor). World of Invention "Summary". Detroit: Thomson Gale. ISBN 0-7876-2759-3.

- "The Accelerated Phaseout of Class I Ozone-Depleting Substances". United States Environmental Protection Agency. 19 August 2010. Retrieved 2015-07-20.

6

Pest Management: An Integrated Study

Pests attack plants and damage them decreasing their yield while also affecting soil quality. Integrated pest management consolidates methods and practices used for profitable pest control. This content elaborates on integrated pest management, biopesticides, insect growth regulator, insecticidal soap and fogger. The section on biopesticides sheds light on biologically benign pesticide alternates that are gaining popularity.

Integrated Pest Management

Integrated pest management (IPM), also known as integrated pest control (IPC) is a broad-based approach that integrates practices for economic control of pests. IPM aims to suppress pest populations below the economic injury level (EIL). The UN's Food and Agriculture Organisation defines IPM as "the careful consideration of all available pest control techniques and subsequent integration of appropriate measures that discourage the development of pest populations and keep pesticides and other interventions to levels that are economically justified and reduce or minimize risks to human health and the environment. IPM emphasizes the growth of a healthy crop with the least possible disruption to agro-ecosystems and encourages natural pest control mechanisms." Entomologists and ecologists have urged the adoption of IPM pest control since the 1970s. IPM allows for safer pest control. This includes managing insects, plant pathogens and weeds.

An IPM boll weevil trap in a cotton field (Manning, South Carolina).

Globalization and increased mobility often allow increasing numbers of invasive species to cross national borders. IPM poses the least risks while maximizing benefits and reducing costs.

For their leadership in developing and spreading IPM worldwide, Perry Adkisson and Ray F. Smith received the 1997 World Food Prize.

History

Shortly after World War II, when synthetic insecticides became widely available, entomologists in California developed the concept of "supervised insect control". Around the same time, entomologists in the US Cotton Belt were advocating a similar approach. Under this scheme, insect control was "supervised" by qualified entomologists and insecticide applications were based on conclusions reached from periodic monitoring of pest and natural-enemy populations. This was viewed as an alternative to calendar-based programs. Supervised control was based on knowledge of the ecology and analysis of projected trends in pest and natural-enemy populations.

Supervised control formed much of the conceptual basis for the "integrated control" that University of California entomologists articulated in the 1950s. Integrated control sought to identify the best mix of chemical and biological controls for a given insect pest. Chemical insecticides were to be used in the manner least disruptive to biological control. The term "integrated" was thus synonymous with "compatible." Chemical controls were to be applied only after regular monitoring indicated that a pest population had reached a level (the economic threshold) that required treatment to prevent the population from reaching a level (the economic injury level) at which economic losses would exceed the cost of the control measures.

IPM extended the concept of integrated control to all classes of pests and was expanded to include all tactics. Controls such as pesticides were to be applied as in integrated control, but these now had to be compatible with tactics for all classes of pests. Other tactics, such as host-plant resistance and cultural manipulations, became part of the IPM framework. IPM combined entomologists, plant pathologists, nematologists and weed scientists.

In the United States, IPM was formulated into national policy in February 1972 when President Richard Nixon directed federal agencies to take steps to advance the application of IPM in all relevant sectors. In 1979, President Jimmy Carter established an interagency IPM Coordinating Committee to ensure development and implementation of IPM practices.

Applications

IPM is used in agriculture, horticulture, human habitations, preventive conservation and general pest control, including structural pest management, turf pest management and ornamental pest management.

Principles

An American IPM system is designed around six basic components:

- Acceptable pest levels—The emphasis is on *control*, not *eradication*. IPM holds that wiping out an entire pest population is often impossible, and the attempt can be expensive and unsafe. IPM programmes first work to establish acceptable pest levels, called action thresholds, and apply controls if those thresholds are crossed. These thresholds are pest and site specific, meaning that it may be acceptable at one site to have a weed such as white clover, but not at another site. Allowing a pest population to survive at a reasonable threshold reduces selection pressure. This lowers the rate at which a pest develops resistance to a control, because if almost all pests are killed then those that have resistance will provide the genetic basis of the future population. Retaining a significant number of unresistant specimens dilutes the prevalence of any resistant genes that appear. Similarly, the repeated use of a single class of controls will create pest populations that are more resistant to that class, whereas alternating among classes helps prevent this.

- Preventive cultural practices—Selecting varieties best for local growing conditions and maintaining healthy crops is the first line of defense. Plant quarantine and 'cultural techniques' such as crop sanitation are next, e.g., removal of diseased plants, and cleaning pruning shears to prevent spread of infections. Beneficial fungi and bacteria are added to the potting media of horticultural crops vulnerable to root diseases, greatly reducing the need for fungicides.

- Monitoring—Regular observation is critically important. Observation is broken into inspection and identification. Visual inspection, insect and spore traps, and other methods are used to monitor pest levels. Record-keeping is essential, as is a thorough knowledge target pest behavior and reproductive cycles. Since insects are cold-blooded, their physical development is dependent on area temperatures. Many insects have had their development cycles modeled in terms of degree-days. The degree days of an environment determines the optimal time for a specific insect outbreak. Plant pathogens follow similar patterns of response to weather and season.

- Mechanical controls—Should a pest reach an unacceptable level, mechanical methods are the first options. They include simple hand-picking, barriers, traps, vacuuming and tillage to disrupt breeding.

- Biological controls—Natural biological processes and materials can provide control, with acceptable environmental impact, and often at lower cost. The main approach is to promote beneficial insects that eat or parasitize target pests. Biological insecticides, derived from naturally occurring microorganisms (e.g.—Bt, entomopathogenic fungi and entomopathogenic nematodes), also fall in this category. Further 'biology-based' or 'ecological' techniques are under evaluation.

- Responsible use—Synthetic pesticides are used as required and often only at specific times in a pest's life cycle. Many newer pesticides are derived from plants or naturally occurring substances (e.g.—nicotine, pyrethrum and insect juvenile hormone analogues), but the toxophore or active component may be altered to provide increased biological activity or stability. Applications of pesticides must reach their intended targets. Matching the application technique to the crop, the pest, and the pesticide is critical. The use of low-volume spray equipment reduces overall pesticide use and labor cost.

An IPM regime can be simple or sophisticated. Historically, the main focus of IPM programmes was on agricultural insect pests. Although originally developed for agricultural pest management, IPM programmes are now developed to encompass diseases, weeds and other pests that interfere with management objectives for sites such as residential and commercial structures, lawn and turf areas, and home and community gardens.

Process

IPM is the selection and use of pest control actions that will ensure favourable economic, ecological and social consequences and is applicable to most agricultural, public health and amenity pest management situations. The IPM process starts with monitoring, which includes inspection and identification, followed by the establishment of economic injury levels. The economic injury levels set the economic threshold level. That is the point when pest damage (and the benefits of treating the pest) exceed the cost of treatment. This can also be an action threshold level for determining an unacceptable level that is not tied to economic injury. Action thresholds are more common in structural pest management and economic injury levels in classic agricultural pest management. An example of an action threshold is one fly in a hospital operating room is not acceptable, but one fly in a pet kennel would be acceptable. Once a threshold has been crossed by the pest population action steps need to be taken to reduce and control the pest. Integrated pest management employ a variety of actions including cultural controls, including physical barriers, biological controls, including adding and conserving natural predators and enemies to the pest, and finally chemical controls or pesticides. Reliance on knowledge, experience, observation and integration of multiple techniques makes IPM appropriate for organic farming (excluding synthetic pesticides). These may or may not include materials listed on the Organic Materials Review Institute (OMRI) Although the pesticides and particularly insecticides used in organic farming and organic gardening are generally safer than synthetic pesticides, they are not always more safe or environmentally friendly than synthetic pesticides and can cause harm. For conventional farms IPM can reduce human and environmental exposure to hazardous chemicals, and potentially lower overall costs.

Risk assessment usually includes four issues: 1) characterization of biological control agents, 2) health risks, 3) environmental risks and 4) efficacy.

Mistaken identification of a pest may result in ineffective actions. E.g., plant damage due to over-watering could be mistaken for fungal infection, since many fungal and viral infections arise under moist conditions.

Monitoring begins immediately, before the pest's activity becomes significant. Monitoring of agricultural pests includes tracking soil/planting media fertility and water quality. Overall plant health and resistance to pests is greatly influenced by pH, alkalinity, of dissolved mineral and Oxygen Reduction Potential. Many diseases are waterborne, spread directly by irrigation water and indirectly by splashing.

Once the pest is known, knowledge of its lifecycle provides the optimal intervention points. For example, weeds reproducing from last year's seed can be prevented with mulches and pre-emergent herbicide.

Pest-tolerant crops such as soybeans may not warrant interventions unless the pests are numerous or rapidly increasing. Intervention is warranted if the expected cost of damage by the pest is more than the cost of control. Health hazards may require intervention that is not warranted by economic considerations.

Specific sites may also have varying requirements. E.g., white clover may be acceptable on the sides of a tee box on a golf course, but unacceptable in the fairway where it could confuse the field of play.

Possible interventions include mechanical/physical, cultural, biological and chemical. Mechanical/physical controls include picking pests off plants, or using netting or other material to exclude pests such as birds from grapes or rodents from structures. Cultural controls include keeping an area free of conducive conditions by removing waste or diseased plants, flooding, sanding, and the use of disease-resistant crop varieties. Biological controls are numerous. They include: conservation of natural predators or augmentation of natural predators, Sterile insect technique (SIT).

Augmentation, inoculative release and inundative release are different methods of biological control that affect the target pest in different ways. Augmentative control includes the periodic introduction of predators. With inundative release, predators are collected, mass-reared and periodically released in large numbers into the pest area. This is used for an immediate reduction in host populations, generally for annual crops, but is not suitable for long run use. With inoculative release a limited number of beneficial organisms are introduced at the start of the growing season. This strategy offers long term control as the organism's progeny affect pest populations throughout the season and is common in orchards. With seasonal inoculative release the beneficials are collected, mass-reared and released seasonally to maintain the beneficial population. This is commonly used in greenhouses. In America and other western countries, inundative releases are predominant, while Asia and the eastern Europe more commonly use inoculation and occasional introductions.

The Sterile insect technique (SIT) is an Area-Wide IPM program that introduces sterile male pests into the pest population to trick females into (unsuccessful) breeding encounters, providing a form of birth control and reducing reproduction rates. The biological controls mentioned above only appropriate in extreme cases, because in the introduction of new species, or supplementation of naturally occurring species can have detrimental ecosystem effects. Biological controls can be used to stop invasive species or pests, but they can become an introduction path for new pests.

Chemical controls include horticultural oils or the application of insecticides and herbicides. A Green Pest Management IPM program uses pesticides derived from plants, such as botanicals, or other naturally occurring materials.

Pesticides can be classified by their modes of action. Rotating among materials with different modes of action minimizes pest resistance.

Evaluation is the process of assessing whether the intervention was effective, whether it produced unacceptable side effects, whether to continue, revise or abandon the program.

Southeast Asia

The Green Revolution of the 1960s and '70s introduced sturdier plants that could support the heavier grain loads resulting from intensive fertilizer use. Pesticide imports by 11 Southeast Asian countries grew nearly sevenfold in value between 1990 and 2010, according to FAO statistics, with disastrous results. Rice farmers become accustomed to spraying soon after planting, triggered by signs of the leaf folder moth, which appears early in the growing season. It causes only superficial damage and doesn't reduce yields. In 1986, Indonesia banned 57 pesticides and completely stopped subsidizing their use. Progress was reversed in the 2000s, when growing production capacity, particularly in China, reduced prices. Rice production in Asia more than doubled. But it left farmers believing more is better—whether it's seed, fertilizer, or pesticides.

The brown planthopper (Nilaparvata lugens), the farmers' main target, has become increasingly resistant. Since 2008, outbreaks have devastated rice harvests throughout Asia, but not in the Mekong Delta. Reduced spraying allowed natural predators to neutralize planthoppers in Vietnam. In 2010 and 2011, massive planthopper outbreaks hit 400,000 hectares of Thai rice fields, causing losses of about $64 million. The Thai government is now pushing the "no spray in the first 40 days" approach.

By contrast early spraying kills frogs, spiders, wasps and dragonflies that prey on the later-arriving and dangerous planthopper and produced resistant strains. Planthoppers now require pesticide doses 500 times greater than originally. Overuse indiscriminately kills beneficial insects and decimates bird and amphibian populations. Pesticides are suspected of harming human health and became a common means for rural Asians to commit suicide.

In 2001, scientists challenged 950 Vietnamese farmers to try IPM. In one plot, each farmer grew rice using their usual amounts of seed and fertilizer, applying pesticide as they chose. In a nearby plot, less seed and fertilizer were used and no pesticides were applied for 40 days after planting. Yields from the experimental plots was as good or better and costs were lower, generating 8% to 10% more net income. The experiment led to the "three reductions, three gains" campaign, claiming that cutting the use of seed, fertilizer and pesticide would boost yield, quality and income. Posters, leaflets, TV commercials and a 2004 radio soap opera that featured a rice farmer who gradually accepted the changes. It didn't hurt that a 2006 planthopper outbreak hit farmers using insecticides harder than those who didn't. Mekong Delta farmers cut insecticide spraying from five times per crop cycle to zero to one.

The Plant Protection Center and the International Rice Research Institute (IRRI) have been encouraging farmers to grow flowers, okra and beans on rice paddy banks, instead of stripping vegetation, as was typical. The plants attract bees and a tiny wasp that eats planthopper eggs, while the vegetables diversify farm incomes.

Agriculture companies offer bundles of pesticides with seeds and fertilizer, with incentives for volume purchases. A proposed law in Vietnam requires licensing pesticide dealers and government approval of advertisements to prevent exaggerated claims. Insecticides that target other pests, such as Scirpophaga incertulas (stem borer), the larvae of moth species that feed on rice plants allegedly yield gains of 21% with proper use.

Biopesticide

Biopesticides, a contraction of 'biological pesticides', include several types of pest management intervention: through predatory, parasitic, or chemical relationships. The term has been associated historically with biological control - and by implication - the manipulation of living organisms. Regulatory positions can be influenced by public perceptions, thus:

- in the EU, biopesticides have been defined as "a form of pesticide based on micro-organisms or natural products".

- the US EPA states that they "include naturally occurring substances that control pests (biochemical pesticides), microorganisms that control pests (microbial pesticides), and pesticidal substances produced by plants containing added genetic material (plant-incorporated protectants) or PIPs".

They are obtained from organisms including plants, bacteria and other microbes, fungi, nematodes, *etc.* They are often important components of integrated pest management (IPM) programmes, and have received much practical attention as substitutes to synthetic chemical plant protection products (PPPs).

Types

Biopesticides can be classified into these classes:

- Microbial pesticides which consist of bacteria, entomopathogenic fungi or viruses (and sometimes includes the metabolites that bacteria or fungi produce). Entomopathogenic nematodes are also often classed as microbial pesticides, even though they are multi-cellular.

- Bio-derived chemicals. Four groups are in commercial use: pyrethrum, rotenone, neem oil, and various essential oils are naturally occurring substances that control (or monitor in the case of pheromones) pests and microbial diseases.

- Plant-incorporated protectants (PIPs) have genetic material from other species incorporated into their genetic material (*i.e.* GM crops). Their use is controversial, especially in many European countries.

- RNAi pesticides, some of which are topical and some of which are absorbed by the crop.

Biopesticides have usually no known function in photosynthesis, growth or other basic aspects of plant physiology. Instead, they are active against biological pests. Many chemical compounds have been identified that are produced by plants to protect them from pests. These materials are biodegradable and renewable alternatives, which can be economical for practical use. Organic farming systems embraces this approach to pest control.

RNAi

RNA interference is under study for possible use as a spray-on insecticide by multiple companies, including Monsanto, Syngenta, and Bayer. Such sprays do not modify the genome of the target plant. The RNA could be modified to maintain its effectiveness as target species evolve tolerance to the original. RNA is a relatively fragile molecule that generally degrades within days or weeks of application. Monsanto estimated costs to be on the order of $5/acre.

RNAi has been used to target weeds that tolerate Monsanto's Roundup herbicide. RNAi mixed with a silicone surfactant that let the RNA molecules enter air-exchange holes in the plant's surface that disrupted the gene for tolerance, affecting it long enough to let the herbicide work. This strategy would allow the continued use of glyphosate-based herbicides, but would not per se assist a herbicide rotation strategy that relied on alternating Roundup with others.

They can be made with enough precision to kill some insect species, while not harming others. Monsanto is also developing an RNA spray to kill potato beetles One challenge is to make it linger on the plant for a week, even if it's raining. The Potato beetle has become resistant to more than 60 conventional insecticides.

Monsanto lobbied the U.S. EPA to exempt RNAi pesticide products from any specific regulations (beyond those that apply to all pesticides) and be exempted from rodent toxicity, allergenicity and residual environmental testing. In 2014 an EPA advisory group found little evidence of a risk to people from eating RNA.

However, in 2012, the Australian Safe Food Foundation alleged that the RNA trigger designed to change wheat's starch content might interfere with the gene for a human liver enzyme. Supporters countered that RNA does not appear to make it past human saliva or stomach acids. The US National Honey Bee Advisory Board told EPA that using RNAi would put natural systems at "the epitome of risk". The beekeepers cautioned that pollinators could be hurt by unintended effects and that the genomes of many insects are still unknown. Other unassessed risks include ecological (given the need for sustained presence for herbicide and other applications) and the possible for RNA drift across species boundaries.

Monsanto has invested in multiple companies for their RNA expertise, including Beeologics (for RNA that kills a parasitic mite that infests hives and for manufacturing technology) and Preceres (nanoparticle lipidoid coatings) and licensed technology from Alnylam and Tekmira. In 2012 Syngenta acquired Devgen, a European RNA partner. Startup Forrest Innovations is investigating RNAi as a solution to citrus greening disease that in 2014 caused 22 percent of oranges in Florida to fall off the trees.

Examples

Bacillus thuringiensis, a bacterial disease of Lepidoptera, Coleoptera and Diptera, is a well-known insecticide example. The toxin from *B. thuringiensis* (Bt toxin) has been incorporated directly into plants through the use of genetic engineering. The use of Bt Toxin is particularly controversial. Its manufacturers claim it has little effect on other organisms, and is more environmentally friendly than synthetic pesticides. However, at least one scientific study has suggested that it may lead to slight histopathological changes on the liver and kidneys of mammals with Bt toxin in their diet.

Other microbial control agents include products based on:

- entomopathogenic fungi (*e.g.Beauveria bassiana,* Paecilomyces fumosoroseus, *Lecanicillium* spp., *Metarhizium* spp.),

- plant disease control agents: include *Trichoderma* spp. and *Ampelomyces quisqualis* (a hyper-parasite of grape powdery mildew); *Bacillus subtilis* is also used to control plant pathogens.

- beneficial nematodes attacking insect (*e.g. Steinernema feltiae*) or slug (*e.g. Phasmarhabditis hermaphrodita*) pests

- entomopathogenic viruses (*e.g.. Cydia pomonella* granulovirus).

- weeds and rodents have also been controlled with microbial agents.

Various naturally occurring materials, including fungal and plant extracts, have been described as biopesticides. Products in this category include:

- Insect pheromones and other semiochemicals

- Fermentation products such as Spinosad (a macro-cyclic lactone)

- Chitosan: a plant in the presence of this product will naturally induce systemic resistance (ISR) to allow the plant to defend itself against disease, pathogens and pests.

- Biopesticides may include natural plant-derived products, which include alkaloids, terpenoids, phenolics and other secondary chemicals. Certain vegetable oils such as canola oil are known to have pesticidal properties. Products based on extracts of plants such as garlic have now been registered in the EU and elsewhere.

Applications

Biopesticides are biological or biologically-derived agents, that are usually applied in a manner similar to chemical pesticides, but achieve pest management in an environmentally friendly way. With all pest management products, but especially microbial agents, effective control requires appropriate formulation and application.

Biopesticides for use against crop diseases have already established themselves on a variety of crops. For example, biopesticides already play an important role in controlling downy mildew diseases. Their benefits include: a 0-Day Pre-Harvest Interval, the ability to use under moderate to severe disease pressure, and the ability to use as a tank mix or in a rotational program with other registered fungicides. Because some market studies estimate that as much as 20% of global fungicide sales are directed at downy mildew diseases, the integration of biofungicides into grape production has substantial benefits in terms of extending the useful life of other fungicides, especially those in the reduced-risk category.

A major growth area for biopesticides is in the area of seed treatments and soil amendments. Fungicidal and biofungicidal seed treatments are used to control soil borne fungal pathogens that cause seed rots, damping-off, root rot and seedling blights. They can also be used to control internal seed–borne fungal pathogens as well as fungal pathogens that are on the surface of the seed. Many biofungicidal products also show capacities to stimulate plant host defence and other physiological processes that can make treated crops more resistant to a variety of biotic and abiotic stresses.

Advantages

- No harmful residues produced, i.e. biodegradable.

- Can be cheaper than chemical pesticides when locally produced.

- Can be more effective than synthetic pesticides in the long-term (as demonstrated, for example, by the LUBILOSA Programme)

Disadvantages

- High specificity: which may require an exact identification of the pest/pathogen and the use of multiple products to be used; although this can also be an advantage in that the biopesticide is less likely to harm species other than the target

- Often slow speed of action (thus making them unsuitable if a pest outbreak is an immediate threat to a crop)

- Often variable efficacy due to the influences of various biotic and abiotic factors (since some biopesticides are living organisms, which bring about pest/pathogen control by multiplying within or nearby the target pest/pathogen)

- Living organisms evolve and increase their resistance to biological, chemical, physical or any other form of control. If the target population is not exterminated or rendered incapable of reproduction, the surviving population can acquire a tolerance of whatever pressures are brought to bear, resulting in an evolutionary arms race.

- Unintended consequences: Studies have found broad spectrum biopesticides have lethal and nonlethal risks for non-target native pollinators such as Melipona quadrifasciata in Brazil.

Insect Growth Regulator

An insect growth regulator (IGR) is a substance (chemical) that inhibits the life cycle of an insect. IGRs are typically used as insecticides to control populations of harmful pests, such as cockroaches or fleas.

Advantages

Many IGRs are labeled "reduced risk" by the Environmental Protection Agency, meaning that they target juvenile harmful insect populations while causing less detrimental effects to beneficial insects. Many beekeepers have reported IGR's negatively affecting brood and young bees . Unlike classic insecticides, IGRs do not affect an insect's nervous system and are thus more worker-friendly within closed environments. IGRs are also more compatible with pest management systems that use biological controls. In addition, while insects can become resistant to insecticides, they are less likely to become resistant to IGRs.

How IGRs Work

As an insect grows, it undergoes a process called molting, where it grows a new exoskeleton under its old one and then sheds to allow the new one to swell to a new size and harden. IGRs prevent an insect from reaching maturity by interfering with the molting process. This in turn curbs infestations because immature insects cannot reproduce. Because IGRs work by interfering with an insect's molting process, they take longer to kill than traditional insecticides. Death typically occurs within 3 to 10 days, depending on the product, the insect's life stage when the product is applied and how quickly the insect develops. Some IGRs cause insects to stop feeding long before they die.

Hormonal IGRs

Hormonal IGRs typically work by mimicking or inhibiting the juvenile hormone (JH), one of the two major hormones involved in insect molting. IGRs can also inhibit the other hormone, ecdysone, large peaks of which trigger the insect to molt. If JH is present at the time of molting, the insect molts into a larger larval form; if absent, it molts into a pupa or adult. IGRs that mimic JH can produce premature molting of young immature stages, disrupting larval development. They can also act on eggs, causing sterility, disrupting behavior or disrupting diapause, the process that causes an insect to become dormant before winter. IGRs that inhibit JH production can cause insects to prematurely molt into a nonfunctional adult. IGRs that inhibit ecdysone can cause pupal mortality by interrupting the transformation of larval tissues into adult tissues during the pupal stage.

Chitin Synthesis Inhibitors

Chitin synthesis inhibitors work by preventing the formation of chitin, a carbohydrate needed to form the insect's exoskeleton. With these inhibitors, an insect grows normally until it molts. The inhibitors prevent the new exoskeleton from forming properly, causing the insect to die. Death may be quick, or take up to several days depending on the insect. Chitin synthesis inhibitors can also kill eggs by disrupting normal embryonic development. Chitin synthesis inhibitors affect insects for longer periods of time than hormonal IGRs. These are also quicker acting but can affect predaceous insects, arthropods and even fish. Compounds include benzoylurea pesticides.

Examples

- Azadirachtin

- Hydroprene (Gentrol)

- Methoprene (Precor)

- Pyriproxyfen (Nyguard, Nylar, Sumilarv)

- Triflumuron (Starycide)

Insecticidal Soap

Insecticidal soap is based on potassium fatty acids and is used to control many plant pests. Because insecticidal soap only works on direct contact with the pests, it is sprayed on plants in way such that the entire plant is wetted. Soaps have a low mammalian toxicity and are therefore considered safe to be used around children and pets and may be used in organic farming.

Composition

Insecticidal soap should be based on long-chain fatty acids (10–18 carbon atoms), because shorter-chain fatty acids tend to be damaging for the plant (phytotoxicity). Short (8-carbon) fatty-acid chains occur for example in coconut oil and palm oil and soaps based on those oils. Recommended concentrations are typically in the range 1–2 percent. One manufacturer recommends a concentration of 0.06% to 0.25% (pure soap equivalent) for most agricultural applications.; another one recommends concentrations of 0.5 to 1% pure soap equivalent. In the European Union, fatty acid potassium salts are registered and allowed as insecticide at a 2% concentration.

Insectidal soap is most effective if it is dissolved in soft water, since the fatty acids in soap tend to precipitate in hard water, thereby reducing the effectivity.

Insecticidal soap is sold commercially for aphid control; these may not always use the word soap, but they will list "potassium salts of fatty acids" or "potassium laurate" as the active ingredient. Certain types of household soaps (not synthetic detergents) are also suitable, but it may be difficult to tell the composition and water content from the label. Potassium-based soaps are typically soft or liquid.

Mechanism of Action

The mechanism of action is not exactly understood. Possible mechanisms are:

- Soap, which enters via the insect's trachea, may disrupt cell membranes, resulting in the cell contents leaking from the damaged cells (cytolysis).

- Soap may dissolve the wax layer on the cuticle ("skin"), which leads to water loss by evaporation.

- Soap may block breathing openings or trachea, which leads to suffocation.

- Soap may interfere with growth hormones.

- Soap may affect insect metabolism.

Affected Organisms

Insecticidal soap works best on soft-bodied insects and arthropods such as aphids, adelgids, mealybugs, spider mites, thrips, jumping plant lice, scale insects, whiteflies, and sawfly larvae. It can also be used for caterpillars and leafhoppers, but these large-bodied insects can be more difficult to control with soaps alone. Many pollinators and predatory insects such as lady beetles, bumblebees, and hoverflies are relatively unaffected. However, soap will kill predatory mites that may help control spider mites. Also, the soft-bodied aphid-eating larvae of lady beetles, lacewing, and hoverflies may be affected negatively. According to one study a single soap application killed about 15% of lacewing and lady-beetle larvae, and about 65% of predatory mites (Amblyseius andersoni).

Green peach aphids are difficult to control since they reproduce quickly (one adult female can deposit up to four nymphs per day) because they tend to reside under the leaves and in leaf axils ("leaf armpits"), where they may not be wetted by a soap spray. Manufacturers indeed state that their insecticidal soaps are only suitable for controlling green peach aphids if used in combination with another insecticide, whereas the same soaps can control other aphids on their own. Among green peach aphids that are in contact with a 2% soap solution, around 95% of the adults and 98% of nymphs die within 48 hours. At 0.75% concentration, the mortality rates are reduced to 75% and 90%, respectively.

Since 2011, insecticidal soap has also been approved in the United States for use against powdery mildew. In the European pesticide registration, its use as an insecticide is listed for aphids, white fly, and spider mites. At different concentrations, it may also be used against algae and moss.

Use

Insecticidal soap solution will only kill pests on contact; it has no residual action against aphids that arrive after it has dried. Therefore, the infested plants must be thoroughly wetted. Repeated applications may be necessary to adequately control high populations of pests.

Soap spray may damage plants, especially at higher concentrations or at temperatures above 32 °C (90 °F). Plant injury may not be apparent until two days after application. Some plant species are particularly sensitive to soap sprays. Highly sensitive plants include: horse chestnut, Japanese maple (Acer), Sorbus aucuparia (mountain ash), Cherimoya fruit, Lamprocapnos (bleeding heart), and sweet pea. Other sensitive plants are, for example: Portulaca, some tomato varieties, Crataegus (hawthorn), cherries, plum, Adiantum (maidenhair fern), Euphorbia milii (crown of thorns), Lantana camara, Tropaeolum (nasturtium), Gardenia jasminoides, Lilium longiflorum (Easter lily). Conifers under (drought) stress or with tender new growth are sensitive as well.

Damage may occur as yellow or brown spotting on the leaves, burned tips, or leaf scorch. Plants under drought stress, young transplants, unrooted cuttings and plants with soft young growth tend to be more sensitive. Sensitivity may be tested on a small portion of a plant or plot before a full-scale application.

One manufacturer recommends that applications are done with 7- to 14-day intervals, with a maximum of three applications, as repeated applications may aggravate phytotoxicity. In addition, water conditioning agents can increase phytotoxicity.

Thanks to its low mammalian toxicity, application of insecticidal soap is typically allowed up to the day of harvest.

Fogger

A fogger is any device that creates a fog, typically containing an insecticide for killing insects and other arthropods. Foggers are often used by consumers as a low cost alternative to professional pest control services. The number of foggers needed for pest control depends on the size of the space to be treated, as stated for safety reasons on the instructions supplied with the devices. The fog may contain flammable gases, leading to a danger of explosion if a fogger is used in a building with a pilot light or other naked flame.

Foggers are also used in aeroponics, a branch of modern agriculture.

Fogger Composition

Total release foggers (TRFs) (also called "bug bombs") are used to kill cockroaches, fleas, and flying insects by filling an area with insecticide. Most foggers contain pyrethroid, pyrethrin, or both as active ingredients. *Pyrethroids* are a class of synthetic insecticides that are chemically similar to natural pyrethrins and have low potential for systemic toxicity in mammals. *Pyrethrins* are insecticides derived from chrysanthemum flowers (pyrethrum). Piperonyl butoxide and n-octyl bicycloheptene dicarboximide often are added to pyrethrin products to inhibit insects' microsomal enzymes that detoxify pyrethrins. To distribute their insecticide, foggers also contain aerosol propellants.

Hazards to Humans

During 2001-2006, a total of 466 fogger-related illnesses or injuries were identified in the United States by the SENSOR-Pesticides program. These illnesses or injuries often resulted from inability or failure to vacate before the fogger discharged, reentry into the treated space too soon after the fogger was discharged, excessive use of foggers for the space being treated, and failure to notify others nearby.

Exposure Symptoms

Pyrethrins have little systemic toxicity in mammals, but they have been reported to induce contact dermatitis, conjunctivitis, and asthma. Signs and symptoms of pyrethroid toxicity include abnormal skin sensation (e.g., burning, itching, tingling, and numbness), dizziness, salivation, headache, fatigue, vomiting, diarrhea, seizure, irritability to sound and touch, and other central nervous system effects.

References

- Charles Perrings; Mark Herbert Williamson; Silvana Dalmazzone (1 January 2000). The Economics of Biological Invasions. Edward Elgar Publishing. ISBN 978-1-84064-378-7.

- "IPM Guidelines". UMassAmherst—Integrated Pest Management, Agriculture and Landscape Program. 2009. Retrieved 13 March 2012.

- Consoli, Fernando L.; Parra, José Roberto Postali; Trichogramma, Roberto Antônio Zucchi (28 September 2010). Egg Parasitoids in Agroecosystems with Emphasis on. Springer. ISBN 978-1-4020-9110-0.

- Rajeev K. Upadhyay; K.G. Mukerji; B. P. Chamola (30 November 2001). Biocontrol Potential and its Exploitation in Sustainable Agriculture: Volume 2: Insect Pests. Springer. pp. 261–. ISBN 978-0-306-46587-1.

- J. C. van Lenteren (2003). Quality Control and Production of Biological Control Agents: Theory and Testing Procedures. CABI. ISBN 978-0-85199-836-7.

- Francis Borgio J, Sahayaraj K and Alper Susurluk I (eds) . Microbial Insecticides: Principles and Applications, Nova Publishers, USA. 492pp. ISBN 978-1-61209-223-2

- "With BioDirect, Monsanto Hopes RNA Sprays Can Someday Deliver Drought Tolerance and Other Traits to Plants on Demand | MIT Technology Review". Retrieved 2015-08-31.

- Centers for Disease Control and Prevention (CDC) (October 17, 2008). "Illnesses and Injuries Related to Total Release Foggers". Morbidity and Mortality Weekly Report. Atlanta, GA: Centers for Disease Control and Prevention. 57 (41): 1125–1129. PMID 18923383. Retrieved November 12, 2008.

Effects of Pesticides

Pesticides adversely affect the environment and the ecosystem balance. They threaten and sometimes decimate species other than the target organisms. This chapter closely considers the down side to pesticide use by drawing together topics like pollinator decline, health effects of pesticides, pesticide poisoning, environmental impact of pesticides and pesticide drift. This chapter discusses the effects of insecticides and pesticides in a critical manner providing key analysis to the subject matter.

Pollinator Decline

The term pollinator decline refers to the reduction in abundance of insect and other animal pollinators in many ecosystems worldwide beginning at the end of the twentieth century, and continuing into the present day.

Pollinators participate in sexual reproduction of many plants, by ensuring cross-pollination, essential for some species, or a major factor in ensuring genetic diversity for others. Since plants are the primary food source for animals, the reduction of one of the primary pollination agents, or even their possible disappearance, has raised concern, and the conservation of pollinators has become part of biodiversity conservation efforts.

Consequences

The value of bee pollination in human nutrition and food for wildlife is immense and difficult to quantify.

60 to 80% of the world's flowering plant species are animal pollinated, and 35% of crop production and 60% of crop plant species depend on animal pollinators. It is commonly said that about one third of human nutrition is due to bee pollination. This includes the majority of fruits, many vegetables (or their seed crop) and secondary effects from legumes such as alfalfa and clover fed to livestock.

In 2000, Drs. Roger Morse and Nicholas Calderone of Cornell University attempted to quantify the effects of just one pollinator, the Western honey bee, on only US food crops. Their calculations came up with a figure of US $14.6 billion in food crop value. In 2009, another study calculated the worldwide value of pollination to agriculture. They calculated the costs using the proportion of each of 100 crops that need pollinators that would not be produced in case insect pollinators disappeared completely. The economic value of insect pollination was then of €153 billion.

Increasing Public Awareness

There are international initiatives (e.g. the International Pollinator Initiative (IPI)) that highlight the need for public participation and awareness of pollinator, such as bees, conservation

Possible Explanations

Pesticide Misuse

Studies have linked neonicotinoid pesticide exposure to bee health decline. These studies add to a growing body of scientific literature and strengthen the case for removing pesticides toxic to bees from the market. Pesticides interfere with honey bee brains, affecting their ability to navigate. Pesticides prevent bumble bees from collecting enough food to produce new queens.

Neonicotinoids are highly toxic to a range of insects, including honey bees and other pollinators. They are taken up by a plant's vascular system and expressed through pollen, nectar and guttation droplets from which bees forage and drink. They are particularly dangerous because, in addition to being acutely toxic in high doses, they also result in serious sub-lethal effects when insects are exposed to chronic low doses, as they are through pollen and water droplets laced with the chemical as well as dust that is released into the air when coated seeds are planted. These effects cause significant problems for the health of individual honey bees as well as the overall health of honey bee colonies and they include disruptions in mobility, navigation, feeding behavior, foraging activity, memory and learning, and overall hive activity.

A French 2012 study of *Apis mellifera* (western honey bee or European honey bee) that focused on the neonicotinoid pesticide thiamethoxam, which is metabolized by bees into clothianidin, a pesticide cited in legal action, tested the hypothesis that a sub-lethal exposure to a neonicotinoid indirectly increases hive death rate through homing failure in foraging honey bees. When exposed to sub-lethal doses of thiamethoxam, at levels present in the environment, honey bees were less likely to return to the hive after foraging than control bees that were tracked with Radio-Frequency Identification (RFID) tagging technology, but not intentionally dosed with pesticides. Higher risks are observed when the homing task is more challenging. The survival rate is even lower when exposed bees are placed in foraging areas with which they are less familiar.

In their 2014 study of *Bombus terrestris* (buff-tailed bumblebee or large earth bumblebee), researchers tracked bees using RFID tagging technology, and found that a sub-lethal exposure to either imidacloprid (a neonicotinoid) and/or a pyrethroid (?-cyhalothrin) over a four-week period caused impairment of the bumble bee's ability to forage.

Imidacloprid effects on bees were examined by researchers exposing colonies of bumble bees to levels of imidacloprid that are realistic in the natural environment, then allowed them to develop under field conditions. Treated colonies had a significantly reduced growth rate and suffered an 85% reduction in production of new queens compared to unexposed control colonies. The study shows that bumble bees, which are wild pollinators, are suffering similar impacts of pesticide exposure to "managed" honey bees. Wild pollinators provide ecosystem services both in agriculture and to a wide range of wild plants that could not survive without insect pollination.

In March, 2012, commercial beekeepers and environmental organizations filed an emergency legal petition with the U.S. Environmental Protection Agency (EPA) to suspend use of clothianidin, urging the agency to adopt safeguards. The legal petition is supported by over one million citizen petition signatures, targets the pesticide for its harmful impacts on honey bees. The petition points to the fact that the EPA failed to follow its own regulations. EPA granted a conditional, or temporary, registration of clothianidin in 2003 without a field study establishing that the pesticide would have no "unreasonable adverse effects" on pollinators. The conditional registration was contingent upon the submission of an acceptable field study, but this requirement has not been met. EPA continues to allow the use of clothianidin nine years after acknowledging that it had an insufficient legal basis for initially allowing its use. Additionally, the product labels on pesticides containing clothianidin are inadequate to prevent excessive damage to non-target organisms, which is a violation of the requirements for using a pesticide and further warrants removing all such mislabeled pesticides from use.

The disappearance of honeybees was documented in the 2009 film *Vanishing of the Bees* by George Langworthy and Maryam Henein.

Rapid Transfer of Parasites and Diseases of Pollinator Species Around The World

Increased international commerce has moved diseases of the honey bee such as American foulbrood and chalkbrood, and parasites such as varroa mites, acarina mites, and the small African hive beetle to new areas of the world, causing much loss of bees in the areas where they do not have much resistance to these pests. Imported fire ants have decimated ground nesting bees in wide areas of the southern US.

Loss of Habitat and Forage

Bees and other pollinators face a higher risk of extinction due to loss of habitat and access to natural food sources. The global dependency on livestock and agriculture has rendered no less than 50% of the earths landmass uninhabitable for bees. The agricultural practice of planting one crop (monoculture) in a given area year after year leads to extreme malnourishment. Regardless if the planted crop does flower and provide food for the bee, the bee will still be malnourished because a single plant cannot meet the nutrient requirements. Furthermore, the crops needed to support livestock (primarily cattle) tend to be grains which do not provide nectar.

Air Pollution

Researchers at the University of Virginia have discovered that air pollution from automobiles and power plants has been inhibiting the ability of pollinators such as bees and butterflies to find the fragrances of flowers. Pollutants such as ozone, hydroxyl, and nitrate radicals bond quickly with volatile scent molecules of flowers, which consequently travel shorter distances intact. There results a vicious cycle in which pollinators travel increasingly longer distances to find flowers providing them nectar, and flowers receive inadequate pollination to reproduce and diversify.

Changes in Seasonal Behaviour Due to Global Warming

In 2014 the Intergovernmental Panel on Climate Change reported that bees, butterflies and other pollinators faced increased risk of extinction because of global warming due to alterations in the seasonal behaviour of species. Climate change was causing bees to emerge at different times in the year when flowering plants were not available.

The Structure of Plant-pollinator Networks

Wild pollinators often visit a large number of plant species and plants are visited by a large number of pollinator species. All these relations together form a network of interactions between plants and pollinators. Surprising similarities were found in the structure of networks consisting out of the interactions between plants and pollinators. This structure was found to be similar in very different ecosystems on different continents, consisting of entirely different species.

The structure of plant-pollinator networks may have large consequences for the way in which pollinator communities respond to increasingly harsh conditions. Mathematical models, examining the consequences of this network structure for the stability of pollinator communities suggest that the specific way in which plant-pollinator networks are organized minimizes competition between pollinators and may even lead to strong indirect facilitation between pollinators when conditions are harsh. This makes that pollinator species together can survive under harsh conditions. But it also means that pollinator species collapse simultaneously when conditions pass a critical point. This simultaneous collapse occurs, because pollinator species depend on each other when surviving under difficult conditions.

Such a community-wide collapse, involving many pollinator species, can occur suddenly when increasingly harsh conditions pass a critical point and recovery from such a collapse might not be easy. The improvement in conditions needed for pollinators to recover, could be substantially larger than the improvement needed to return to conditions at which the pollinator community collapsed.

Solutions

Conservation and Restoration

Efforts are being made to sustain pollinator diversity in agro and natural eco-systems by some environmental groups. Prairie restoration, establishment of wildlife preserves, and encouragement of diverse wildlife landscaping rather than mono culture lawns, are examples of ways to help pollinators.

In June 2014 the Obama administration published a fact sheet "The Economic Challenge Posed by Declining Pollinator Populations", which stated that "President's 2015 Budget recommends approximately $50 million across multiple agencies within USDA to [...] strengthen pollinator habitat in core areas, double the number of acres in the Conservation Reserve Program that are dedicated to pollinator health [...]".

Research

The Obama administration's 2015 Budget also recommended to "enhance research at USDA and through public-private grants, [...] and increase funding for surveys to determine the impacts on pollinator losses".

SmartBees is a European research project of 16 entities (universities, research institutions and companies) funded by the EU, headquartered in Berlin. Its goal is to elicit causes of resistance to CCD, develop breeding to increase CCD resistance and to counteract the replacement of many native European bees with only two specific races.

CoLOSS (Prevention of honey bee COlony LOSSes) is an international, non-profit association headquartered in Bern, Switzerland to "improve the well-being of bees at a global level", composed of researchers, veterinarians, agriculture extension specialists, and students from 69 countries. Their 3 core projects are standardization of methods for studying the honey bee, colony loss monitoring and B-RAP (Bridging Research and Practice).

Contract Pollination

The decline of pollinators is compensated to some extent by beekeepers becoming migratory, following the bloom northward in the spring from southern wintering locations. Migration may be for traditional honey crops, but increasingly is for contract pollination to supply the needs for growers of crops that require it.

Health Effects of Pesticides

Health effects of pesticides may be acute or delayed in those who are exposed. A 2007 systematic review found that "most studies on non-Hodgkin lymphoma and leukemia showed positive associations with pesticide exposure" and thus concluded that cosmetic use of pesticides should be decreased. Strong evidence also exists for other negative outcomes from pesticide exposure including neurological problems, birth defects, fetal death, and neurodevelopmental disorder.

According to The Stockholm Convention on Persistent Organic Pollutants, 9 of the 12 most dangerous and persistent chemicals are pesticides.

Acute Effects

Acute health problems may occur in workers that handle pesticides, such as abdominal pain, dizziness, headaches, nausea, vomiting, as well as skin and eye problems. In China, an estimated half million people are poisoned by pesticides each year, 500 of whom die. Pyrethrins, insecticides commonly used in common bug killers, can cause a potentially deadly condition if breathed in.

Long-term Effects

Cancer

Many studies have examined the effects of pesticide exposure on the risk of cancer. Associations have been found with: leukemia, lymphoma, brain, kidney, breast, prostate, pancreas, liver, lung, and skin cancers. This increased risk occurs with both residential and occupational exposures. Increased rates of cancer have been found among farm workers who apply these chemicals. A mother's occupational exposure to pesticides during pregnancy is associated with an increases in her child's risk of leukemia, Wilms' tumor, and brain cancer. Exposure to insecticides within the home and herbicides outside is associated with blood cancers in children.

Neurological

Evidence links pesticide exposure to worsened neurological outcomes. The risk of developing Parkinson's disease is 70% greater in those exposed to even low levels of pesticides. People with Parkinson's were 61% more likely to report direct pesticide application than were healthy relatives. Both insecticides and herbicides significantly increased the risk of Parkinson's disease. There are also concerns that long-term exposures may increase the risk of dementia.

The United States Environmental Protection Agency finished a 10-year review of the organophosphate pesticides following the 1996 Food Quality Protection Act, but did little to account for developmental neurotoxic effects, drawing strong criticism from within the agency and from outside researchers. Comparable studies have not been done with newer pesticides that are replacing organophosphates.

Reproductive Effects

Strong evidence links pesticide exposure to birth defects, fetal death and altered fetal growth. In the United States, increase in birth defects is associated with conceiving in the same period of the year when agrochemicals are in elevated concentrations in surface water. Agent Orange, a 50:50 mixture of 2,4,5-T and 2,4-D, has been associated with bad health and genetic effects in Malaya and Vietnam. It was also found that offspring that were at some point exposed to pesticides had a low birth weight and had developmental defects.

Fertility

A number of pesticides including dibromochlorophane and 2,4-D has been associated with impaired fertility in males. Pesticide exposure resulted in reduced fertility in males, genetic alterations in sperm, a reduced number of sperm, damage to germinal epithelium and altered hormone function.

Other

Some studies have found increased risks of dermatitis in those exposed.

Additionally, studies have indicated that pesticide exposure is associated with long-term health problems such as respiratory problems, including asthma, memory disorders and depression. Summaries of peer-reviewed research have examined the link between pesticide exposure and neurologic outcomes and cancer, perhaps the two most significant things resulting in organophosphate-exposed workers.

According to researchers from the National Institutes of Health (NIH), licensed pesticide applicators who used chlorinated pesticides on more than 100 days in their lifetime were at greater risk of diabetes. One study found that associations between specific pesticides and incident diabetes ranged from a 20 percent to a 200 percent increase in risk. New cases of diabetes were reported by 3.4 percent of those in the lowest pesticide use category compared with 4.6 percent of those in the highest category. Risks were greater when users of specific pesticides were compared with applicators who never applied that chemical.

Route of Exposure

People can be exposed to pesticides by a number of different routes including: occupation, in the home, at school and in their food.

There are concerns that pesticides used to control pests on food crops are dangerous to people who consume those foods. These concerns are one reason for the organic food movement. Many food crops, including fruits and vegetables, contain pesticide residues after being washed or peeled. Chemicals that are no longer used but that are resistant to breakdown for long periods may remain in soil and water and thus in food.

The United Nations Codex Alimentarius Commission has recommended international standards for maximum residue limits (MRLs), for individual pesticides in food.

In the EU, MRLs are set by DG-SANCO.

In the United States, levels of residues that remain on foods are limited to tolerance levels that are established by the U.S. Environmental Protection Agency and are considered safe. The EPA sets the tolerances based on the toxicity of the pesticide and its breakdown products, the amount and frequency of pesticide application, and how much of the pesticide (i.e., the residue) remains in or on food by the time it is marketed and prepared. Tolerance levels are obtained using scientific risk assessments that pesticide manufacturers are required to produce by conducting toxicological studies, exposure modeling and residue studies before a particular pesticide can be registered, however, the effects are tested for single pesticides, and there is little information on possible synergistic effects of exposure to multiple pesticide traces in the air, food and water.

Strawberries and tomatoes are the two crops with the most intensive use of soil fumigants. They are particularly vulnerable to several type of diseases, insects, mites, and parasitic worms. In 2003, in California alone, 3.7 million pounds (1,700 metric tons) of metham sodium were used on tomatoes. In recent years other farmers have demonstrated that it is possible to produce strawberries and tomatoes without the use of harmful chemicals and in a cost-effective way.

Exposure routes other than consuming food that contains residues, in particular pesticide drift, are potentially significant to the general public.

Some pesticides can remain in the environment for prolonged periods of time. For example, most people in the United States still have detectable levels of DDT in their bodies even though it was banned in the US in 1972.

Prevention

Pesticides exposure cannot be studied in placebo controlled trials as this would be unethical. A definitive cause effect relationship therefore cannot be established. Consistent evidence can and has been gathered through other study designs. The precautionary principle is thus frequently used in environmental law such that absolute proof is not required before efforts to decrease exposure to potential toxins are enacted.

The American Medical Association recommend limiting exposure to pesticides. They came to this conclusion due to the fact that surveillance systems currently in place are inadequate to determine problems related to exposure. The utility of applicator certification and public notification programs are also of unknown value in their ability to prevent adverse outcomes.

Epidemiology

The World Health Organization and the UN Environment Programme estimate that each year, 3 million workers in agriculture in the developing world experience severe poisoning from pesticides, about 18,000 of whom die. According to one study, as many as 25 million workers in developing countries may suffer mild pesticide poisoning yearly. Detectable levels of 50 different pesticides were found in the blood of a representative sample of the U.S. population.

Society and Culture

Concerns regarding conflict of interests regarding the research base have been raised. A number of researchers involved with pesticides have been found to have undisclosed ties to industry including: Richard Doll or the Imperial Cancer Research Fund in England and Hans-Olov Adami of the Karolinska Institute in Sweden.

Other Animals

A number of pesticides including clothianidin, dinotefuran, imidacloprid are toxic to bees. Exposure to pesticides may be one of the contributory factors to colony collapse disorder. A study in North Carolina indicated that more than 30 percent of the quail tested were made sick by one aerial insecticide application. Once sick, wild birds may neglect their young, abandon their nests, and become more susceptible to predators or disease.

Pesticide Poisoning

A pesticide poisoning occurs when chemicals intended to control a pest affect non-target organisms such as humans, wildlife, or bees. There are three types of pesticide poisoning. The first of the three is a single and short-term very high level of exposure which can be experienced by individuals who commit suicide, as well as pesticide formulators. The second type of poisoning is long-term high-level exposure, which can occur in pesticide formulators and manufacturers. The third type of poisoning is a long-term low-level exposure, which individuals are exposed to from sources such as pesticide residues in food as well as contact with pesticide residues in the air, water, soil, sediment, food materials, plants and animals.

In developing countries, such as Sri Lanka, pesticide poisonings from short-term very high level of exposure (acute poisoning) is the most worrisome type of poisoning. However, in developed countries, such as Canada, it is the complete opposite: acute pesticide poisoning is controlled, thus making the main issue long-term low-level exposure of pesticides.

Cause

The most common exposure scenarios for pesticide-poisoning cases are accidental or suicidal poisonings, occupational exposure, by-stander exposure to off-target drift, and the general public who are exposed through environmental contamination.

Accidental and Suicidal

Self-poisoning with agricultural pesticides represents a major hidden public health problem accounting for approximately one-third of all suicides worldwide. It is one of the most common forms of self-injury in the Global South. The World Health Organization estimates that 300,000 people die from self-harm each year in the Asia-Pacific region alone. Most cases of intentional pesticide poisoning appear to be impulsive acts undertaken during stressful events, and the availability of pesticides strongly influences the incidence of self poisoning. Pesticides are the agents most frequently used by farmers and students in India to commit suicide.

Occupational

Pesticide poisoning is an important occupational health issue because pesticides are used in a large number of industries, which puts many different categories of workers at risk. Extensive use puts agricultural workers in particular at increased risk for pesticide illnesses. Workers in other industries are at risk for exposure as well. For example, commercial availability of pesticides in stores puts retail workers at risk for exposure and illness when they handle pesticide products. The ubiquity of pesticides puts emergency responders such as fire-fighters and police officers at risk, because they are often the first responders to emergency events and may be unaware of the presence of a poisoning hazard. The process of aircraft disinsection, in which pesticides are used on inbound international flights for insect and disease control, can also make flight attendants sick.

Different job functions can lead to different levels of exposure. Most occupational exposures are caused by absorption through exposed skin such as the face, hands, forearms, neck, and chest. This exposure is sometimes enhanced by inhalation in settings including spraying operations in greenhouses and other closed environments, tractor cabs, and the operation of rotary fan mist sprayers.

Residential

When thinking of pesticide poisoning, one does not take into consideration the contribution that is made of their own household. The majority of households in Canada use pesticides while taking part in activities such as gardening. In Canada 96 percent of households report having a lawn or a garden. 56 percent of the households who have a lawn or a garden utilize fertilizer or pesticide. This form of pesticide use may contribute to the third type of poisoning, which is caused by long-term low-level exposure. As mentioned before, long-term low-level exposure affects individuals from sources such as pesticide residues in food as well as contact with pesticide residues in the air, water, soil, sediment, food materials, plants and animals.

Pathophysiology

Organochlorines

The organochlorine pesticides, like DDT, aldrin, and dieldrin are extremely persistent and accumulate in fatty tissue. Through the process of bioaccumulation (lower amounts in the environment get magnified sequentially up the food chain), large amounts of organochlorines can accumulate in top species like humans. There is substantial evidence to suggest that DDT, and its metabolite DDE, act as endocrine disruptors, interfering with hormonal function of estrogen, testosterone, and other steroid hormones.

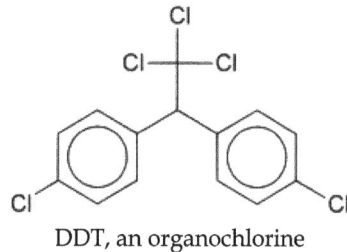

DDT, an organochlorine

Anticholinesterase Compounds

Certain organophosphates have long been known to cause a delayed-onset toxicity to nerve cells, which is often irreversible. Several studies have shown persistent deficits in cognitive function in workers chronically exposed to pesticides. Newer evidence suggests that these pesticides may cause developmental neurotoxicity at much lower doses and without depression of plasma cholinesterase levels.

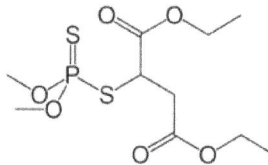

Malathion, an organophosphate anticholinesterase

Diagnosis

Most pesticide-related illnesses have signs and symptoms that are similar to common medical conditions, so a complete and detailed environmental and occupational history is essential for correctly diagnosing a pesticide poisoning. A few additional screening questions about the patient's work and home environment, in addition to a typical health questionnaire, can indicate whether there was a potential pesticide poisoning.

If one is regularly using carbamate and organophosphate pesticides, it is important to obtain a baseline cholinesterase test. Cholinesterase is an important enzyme of the nervous system, and these chemical groups kill pests and potentially injure or kill humans by inhibiting cholinesterase. If one has had a baseline test and later suspects a poisoning, one can identify the extent of the problem by comparison of the current cholinesterase level with the baseline level.

Prevention

Accidental poisonings can be avoided by proper labeling and storage of containers. When handling or applying pesticides, exposure can be significantly reduced by protecting certain parts of the body where the skin shows increased absorption, such as the scrotal region, underarms, face, scalp, and hands. Using chemical-resistant gloves has been shown to reduce contamination by 33–86%.

Further methods in order to aid prevention of acute pesticide poisoning, concerning both accidental death and suicides, there could be a method for national governments to control accessibility. The pesticides most toxic to humans if restricted has the possibility to reduce deaths. There could also be designated locations in rural living areas and cities used to safely store toxic pesticides in order to gain control over usage.

Treatment

Specific treatments for acute pesticide poisoning are often dependent on the pesticide or class of pesticide responsible for the poisoning. However, there are basic management techniques that are applicable to most acute poisonings, including skin decontamination, airway protection, gastrointestinal decontamination, and seizure treatment.

Decontamination of the skin is performed while other life-saving measures are taking place. Clothing is removed, the patient is showered with soap and water, and the hair is shampooed to remove chemicals from the skin and hair. The eyes are flushed with water for 10–15 minutes. The patient is intubated and oxygen administered, if necessary. In more severe cases, pulmonary ventilation must sometimes be supported mechanically. Seizures are typically managed with lorazepam, phenytoin and phenobarbitol, or diazepam (particularly for organochlorine poisonings).

Gastric lavage is not recommended to be used routinely in pesticide poisoning management, as clinical benefit has not been confirmed in controlled studies; it is indicated only when the patient has ingested a potentially life-threatening amount of poison and presents within 60 minutes of ingestion. An orogastric tube is inserted and the stomach is flushed with saline to try to remove the poison. If the patient is neurologically impaired, a cuffed endotracheal tube inserted beforehand for airway protection. Studies of poison recovery at 60 minutes have shown recovery of 8%–32%. However, there is also evidence that lavage may flush the material into the small intestine, increasing absorption. Lavage is contra-indicated in cases of hydrocarbon ingestion.

Activated charcoal is sometimes administered as it has been shown to be successful with some pesticides. Studies have shown that it can reduce the amount absorbed if given within 60 minutes, though there is not enough data to determine if it is effective if time from ingestion is prolonged. Syrup of ipecac is no longer recommended for most pesticide poisonings.

Urinary alkalinisation has been used in acute poisonings from chlorophenoxy herbicides (such as 2,4-D, MCPA, 2,4,5-T and mecoprop); however, evidence to support its use is poor.

Epidemiology

Acute pesticide poisoning is a large-scale problem, especially in developing countries.

"Most estimates concerning the extent of acute pesticide poisoning have been based on data from hospital admissions which would include only the more serious cases. The latest estimate by a WHO task group indicates that there may be 1 million serious unintentional poisonings each year and in addition 2 million people hospitalized for suicide attempts with pesticides. This necessarily reflects only a fraction of the real problem. On the basis of a survey of self-reported minor poisoning carried out in the Asian region, it is estimated that there could be as many as 25 million agricultural workers in the developing world suffering an episode of poisoning each year." In Canada in 2007 more than 6000 cases of acute pesticide poisoning occurred.

Estimating the numbers of chronic poisonings worldwide is more difficult.

Society and Culture

Rachel Carson's *Silent Spring* brought about the first major wave of public concern over the chronic effects of pesticides.

In Other Animals

An obvious side effect of using a chemical meant to kill is that one is likely to kill more than just the desired organism. Contact with a sprayed plant or "weed" can have an effect upon local wildlife, most notably insects. A cause for concern is how pests, the reason for pesticide use, are building up a resistance. Phytophagous insects are able to build up this resistance because they are easily capable of evolutionary diversification and adaption. The problem this presents is that in order to obtain the same desired effect of the pesticides they have to be made increasingly stronger as time goes on. Repercussions of the use of stronger pesticides on vegetation has a negative result on the surrounding environment, but also would contribute to consumers' long-term low-level exposure.

Environmental Impact of Pesticides

The environmental impact of pesticides consists of the effects of pesticides on non-target species. Over 98% of sprayed insecticides and 95% of herbicides reach a destination other than their target species, because they are sprayed or spread across entire agricultural fields. Runoff can carry pesticides into aquatic environments while wind can carry them to other fields, grazing areas, human settlements and undeveloped areas, potentially affecting other species. Other problems emerge from poor production, transport and storage practices. Over time, repeated application increases pest resistance, while its effects on other species can facilitate the pest's resurgence.

Each pesticide or pesticide class comes with a specific set of environmental concerns. Such undesirable effects have led many pesticides to be banned, while regulations have limited and/or reduced the use of others. Over time, pesticides have generally become less persistent and more species-specific, reducing their environmental footprint. In addition the amounts of pesticides applied per hectare have declined, in some cases by 99%. However, the global spread of pesticide use, including the use of older/obsolete pesticides that have been banned in some jurisdictions, has increased overall.

Agriculture and the Environment

The arrival of humans in an area, to live or to conduct agriculture, necessarily has environmental impacts. These range from simple crowding out of wild plants in favor of more desirable cultivars to larger scale impacts such as reducing biodiversity by reducing food availability of native species, which can propagate across food chains. The use of agricultural chemicals such as fertilizer and pesticides magnify those impacts. While advances in agrochemistry have reduced those impacts, for example by the replacement of long-lived chemicals with those that reliably degrade, even in the best case they remain substantial. These effects are magnified by the use of older chemistries and poor management practices.

History

While concern ecotoxicology began with acute poisoning events in the late 19th century; public concern over the undesirable environmental effects of chemicals arose in the early 1960s with the publication of Rachel Carson's book, *Silent Spring*. Shortly thereafter, DDT, originally used to combat malaria, and its metabolites were shown to cause population-level effects in raptorial birds. Initial studies in industrialized countries focused on acute mortality effects mostly involving birds or fish.

Data on pesticide usage remain scattered and/or not publicly available (3). The common practice of incident registration is inadequate for understanding the entirety of effects.

Since 1990, research interest has shifted from documenting incidents and quantifying chemical exposure to studies aimed at linking laboratory, mesocosm and field experiments. The proportion of effect-related publications has increased. Animal studies mostly focus on fish, insects, birds, amphibians and arachnids.

Since 1993, the United States and the European Union have updated pesticide risk assessments, ending the use of acutely toxic organophosphate and carbamate insecticides. Newer pesticides aim at efficiency in target and minimum side effects in nontarget organisms. The phylogenetic proximity of beneficial and pest species complicates the project.

One of the major challenges is to link the results from cellular studies through many levels of increasing complexity to ecosystems.

Specific Pesticide Effects

Pesticide environmental effects

Pesticide/class	Effect(s)
Organochlorine DDT/DDE	Egg shell thinning in raptorial birds
	Endocrine disruptor
	Thyroid disruption properties in rodents, birds, amphibians and fish
	Acute mortality attributed to inhibition of acetylcholine esterase activity
:DDT	Carcinogen
	Endocrine disruptor
:DDT/Diclofol, Dieldrin and Toxaphene	Juvenile population decline and adult mortality in wildlife reptiles
:DDT/Toxaphene/Parathion	Susceptibility to fungal infection
Triazine	Earthworms became infected with monocystid gregarines
:Chlordane	Interact with vertebrate immune systems
Carbamates, the phenoxy herbicide 2,4-D, and atrazine	Interact with vertebrate immune systems

Anticholinesterase	Bird poisoning
	Animal infections, disease outbreaks and higher mortality.
Organophosphate	Thyroid disruption properties in rodents, birds, amphibians and fish
	Acute mortality attributed to inhibition of acetylcholine esterase activity
	Immunotoxicity, primarily caused by the inhibition of serine hydrolases or esterases
	Oxidative damage
	Modulation of signal transduction pathways
	Impaired metabolic functions such as thermoregulation, water and/or food intake and behavior, impaired development, reduced reproduction and hatching success in vertebrates.
Carbamate	Thyroid disruption properties in rodents, birds, amphibians and fish
	Impaired metabolic functions such as thermoregulation, water and/or food intake and behavior, impaired development, reduced reproduction and hatching success in vertebrates.
	Interact with vertebrate immune systems
	Acute mortality attributed to inhibition of acetylcholine esterase activity
Phenoxy herbicide 2,4-D	Interact with vertebrate immune systems
Atrazine	Interact with vertebrate immune systems
	Reduced northern leopard frog (Rana pipiens) populations because atrazine killed phytoplankton, thus allowing light to penetrate the water column and periphyton to assimilate nutrients released from the plankton. Periphyton growth provided more food to grazers, increasing snail populations, which provide intermediate hosts for trematode.
Pyrethroid	Thyroid disruption properties in rodents, birds, amphibians and fish
Thiocarbamate	Thyroid disruption properties in rodents, birds, amphibians and fish
Triazine	Thyroid disruption properties in rodents, birds, amphibians and fish
Triazole	Thyroid disruption properties in rodents, birds, amphibians and fish
	Impaired metabolic functions such as thermoregulation, water and/or food intake and behavior, impaired development, reduced reproduction and hatching success in vertebrates.
Nicotinoid	respiratory, cardiovascular, neurological, and immunological toxicity in rats and humans
	Disrupt biogenic amine signaling and cause subsequent olfactory dysfunction, as well as affecting foraging behavior, learning and memory.
Imidacloprid, Imidacloprid/ pyrethroid λ-cyhalothrin	Impaired foraging, brood development, and colony success in terms of growth rate and new queen production.
Thiamethoxa	High honey bee worker mortality due to homing failure (risks for colony collapse remain controversial)
Spinosyns	Affect various physiological and behavioral traits of beneficial arthropods, particularly hymenopterans
Bt corn/Cry	Reduced abundance of some insect taxa, predominantly susceptible Lepidopteran herbivores as well as their predators and parasitoids.
Herbicide	Reduced food availability and adverse secondary effects on soil invertebrates and butterflies
	Decreased species abundance and diversity in small mammals.

Benomyl	Altered the patch-level floral display and later a two-thirds reduction of the total number of bee visits and in a shift in the visitors from large-bodied bees to small-bodied bees and flies
Herbicide and planting cycles	Reduced survival and reproductive rates in seed-eating or carnivorous birds

Air

Pesticides can contribute to air pollution. Pesticide drift occurs when pesticides suspended in the air as particles are carried by wind to other areas, potentially contaminating them. Pesticides that are applied to crops can volatilize and may be blown by winds into nearby areas, potentially posing a threat to wildlife. Weather conditions at the time of application as well as temperature and relative humidity change the spread of the pesticide in the air. As wind velocity increases so does the spray drift and exposure. Low relative humidity and high temperature result in more spray evaporating. The amount of inhalable pesticides in the outdoor environment is therefore often dependent on the season. Also, droplets of sprayed pesticides or particles from pesticides applied as dusts may travel on the wind to other areas, or pesticides may adhere to particles that blow in the wind, such as dust particles. Ground spraying produces less pesticide drift than aerial spraying does. Farmers can employ a buffer zone around their crop, consisting of empty land or non-crop plants such as evergreen trees to serve as windbreaks and absorb the pesticides, preventing drift into other areas. Such windbreaks are legally required in the Netherlands.

Spraying a mosquito pesticide over a city

Pesticides that are sprayed on to fields and used to fumigate soil can give off chemicals called volatile organic compounds, which can react with other chemicals and form a pollutant called tropospheric ozone. Pesticide use accounts for about 6 percent of total tropospheric ozone levels.

Water

In the United States, pesticides were found to pollute every stream and over 90% of wells sampled in a study by the US Geological Survey. Pesticide residues have also been found in rain and groundwater. Studies by the UK government showed that pesticide concentrations exceeded those allowable for drinking water in some samples of river water and groundwater.

Pesticide impacts on aquatic systems are often studied using a hydrology transport model to study movement and fate of chemicals in rivers and streams. As early as the 1970s quantitative analysis of pesticide runoff was conducted in order to predict amounts of pesticide that would reach surface waters.

There are four major routes through which pesticides reach the water: it may drift outside of the intended area when it is sprayed, it may percolate, or leach, through the soil, it may be carried to the water as runoff, or it may be spilled, for example accidentally or through neglect. They may also be carried to water by eroding soil. Factors that affect a pesticide's ability to contaminate water include its water solubility, the distance from an application site to a body of water, weather, soil type, presence of a growing crop, and the method used to apply the chemical.

Maximum limits of allowable concentrations for individual pesticides in public bodies of water are set by the Environmental Protection Agency in the US. Similarly, the government of the United Kingdom sets Environmental Quality Standards (EQS), or maximum allowable concentrations of some pesticides in bodies of water above which toxicity may occur. The European Union also regulates maximum concentrations of pesticides in water.

Soil

Many of the chemicals used in pesticides are persistent soil contaminants, whose impact may endure for decades and adversely affect soil conservation.

Caution against entering a field sprayed with sulphuric acid

The use of pesticides decreases the general biodiversity in the soil. Not using the chemicals results in higher soil quality, with the additional effect that more organic matter in the soil allows for higher water retention. This helps increase yields for farms in drought years, when organic farms have had yields 20-40% higher than their conventional counterparts. A smaller content of organic matter in the soil increases the amount of pesticide that will leave the area of application, because organic matter binds to and helps break down pesticides.

Degradation and sorption are both factors which influence the persistence of pesticides in soil. Depending on the chemical nature of the pesticide, such processes control directly the transportation from soil to water, and in turn to air and our food. Breaking down organic substances, degradation, involves interactions among microorganisms in the soil. Sorption affects bioaccumulation of pesticides which are dependent on organic matter in the soil. Weak organic acids have been shown to be weakly sorbed by soil, because of pH and mostly acidic structure. Sorbed chemicals have been shown to be less accessible to microorganisms. Aging mechanisms are poorly understood but as residence times in soil increase, pesticide residues become more resistant to degradation and extraction as they lose biological activity.

Effect on Plants

Nitrogen fixation, which is required for the growth of higher plants, is hindered by pesticides in soil. The insecticides DDT, methyl parathion, and especially pentachlorophenol have been shown to interfere with legume-rhizobium chemical signaling. Reduction of this symbiotic chemical signaling results in reduced nitrogen fixation and thus reduced crop yields. Root nodule formation in these plants saves the world economy $10 billion in synthetic nitrogen fertilizer every year.

Crop spraying

Pesticides can kill bees and are strongly implicated in pollinator decline, the loss of species that pollinate plants, including through the mechanism of Colony Collapse Disorder, in which worker bees from a beehive or western honey bee colony abruptly disappear. Application of pesticides to crops that are in bloom can kill honeybees, which act as pollinators. The USDA and USFWS estimate that US farmers lose at least $200 million a year from reduced crop pollination because pesticides applied to fields eliminate about a fifth of honeybee colonies in the US and harm an additional 15%.

On the other side, pesticides have some direct harmful effect on plant including poor root hair development, shoot yellowing and reduced plant growth.

Effect on Animals

Many kinds of animals are harmed by pesticides, leading many countries to regulate pesticide usage through Biodiversity Action Plans.

Animals including humans may be poisoned by pesticide residues that remain on food, for example when wild animals enter sprayed fields or nearby areas shortly after spraying.

Pesticides can eliminate some animals' essential food sources, causing the animals to relocate, change their diet or starve. Residues can travel up the food chain; for example, birds can be harmed when they eat insects and worms that have consumed pesticides. Earthworms digest organic matter and increase nutrient content in the top layer of soil. They protect human health by ingesting decomposing litter and serving as bioindicators of soil activity. Pesticides have had harmful effects on growth and reproduction on earthworms. Some pesticides can bioaccumulate, or build up to toxic levels in the bodies of organisms that consume them over time, a phenomenon that impacts species high on the food chain especially hard.

Birds

The US Fish and Wildlife Service estimates that 72 million birds are killed by pesticides in the United States each year. Bald eagles are common examples of nontarget organisms that are impacted by pesticide use. Rachel Carson's book *Silent Spring* dealt with damage to bird species due to pesticide bioaccumulation. There is evidence that birds are continuing to be harmed by pesticide use. In the farmland of the United Kingdom, populations of ten different bird species declined by 10 million breeding individuals between 1979 and 1999, allegedly from loss of plant and invertebrate species on which the birds feed. Throughout Europe, 116 species of birds were threatened as of 1999. Reductions in bird populations have been found to be associated with times and areas in which pesticides are used. DDE-induced egg shell thinning has especially affected European and North American bird populations. In another example, some types of fungicides used in peanut farming are only slightly toxic to birds and mammals, but may kill earthworms, which can in turn reduce populations of the birds and mammals that feed on them.

In England, the use of pesticides in gardens and farmland has seen a reduction in the number of common chaffinches

Some pesticides come in granular form. Wildlife may eat the granules, mistaking them for grains of food. A few granules of a pesticide may be enough to kill a small bird.

The herbicide paraquat, when sprayed onto bird eggs, causes growth abnormalities in embryos and reduces the number of chicks that hatch successfully, but most herbicides do not directly cause much harm to birds. Herbicides may endanger bird populations by reducing their habitat.

Aquatic Life

Fish and other aquatic biota may be harmed by pesticide-contaminated water. Pesticide surface runoff into rivers and streams can be highly lethal to aquatic life, sometimes killing all the fish in a particular stream.

Using an aquatic herbicide

Application of herbicides to bodies of water can cause fish kills when the dead plants decay and consume the water's oxygen, suffocating the fish. Herbicides such as copper sulfite that are applied to water to kill plants are toxic to fish and other water animals at concentrations similar to those used to kill the plants. Repeated exposure to sublethal doses of some pesticides can cause physiological and behavioral changes that reduce fish populations, such as abandonment of nests and broods, decreased immunity to disease and decreased predator avoidance.

Wide field margins can reduce fertilizer and pesticide pollution in streams and rivers

Application of herbicides to bodies of water can kill plants on which fish depend for their habitat.

Pesticides can accumulate in bodies of water to levels that kill off zooplankton, the main source of food for young fish. Pesticides can also kill off insects on which some fish feed, causing the fish to travel farther in search of food and exposing them to greater risk from predators.

The faster a given pesticide breaks down in the environment, the less threat it poses to aquatic life. Insecticides are typically more toxic to aquatic life than herbicides and fungicides.

Amphibians

In the past several decades, amphibian populations have declined across the world, for unexplained reasons which are thought to be varied but of which pesticides may be a part.

Pesticide mixtures appear to have a cumulative toxic effect on frogs. Tadpoles from ponds containing multiple pesticides take longer to metamorphose and are smaller when they do, decreasing their ability to catch prey and avoid predators. Exposing tadpoles to the organochloride endosulfan at levels likely to be found in habitats near fields sprayed with the chemical kills the tadpoles and causes behavioral and growth abnormalities.

The herbicide atrazine can turn male frogs into hermaphrodites, decreasing their ability to reproduce. Both reproductive and nonreproductive effects in aquatic reptiles and amphibians have been reported. Crocodiles, many turtle species and some lizards lack sex-distinct chromosomes until after fertilization during organogenesis, depending on temperature. Embryonic exposure in turtles to various PCBs causes a sex reversal. Across the United States and Canada disorders such as decreased hatching success, feminization, skin lesions, and other developmental abnormalities have been reported.

Humans

Pesticides can enter the body through inhalation of aerosols, dust and vapor that contain pesticides; through oral exposure by consuming food/water; and through skin exposure by direct contact. Pesticides secrete into soils and groundwater which can end up in drinking water, and pesticide spray can drift and pollute the air.

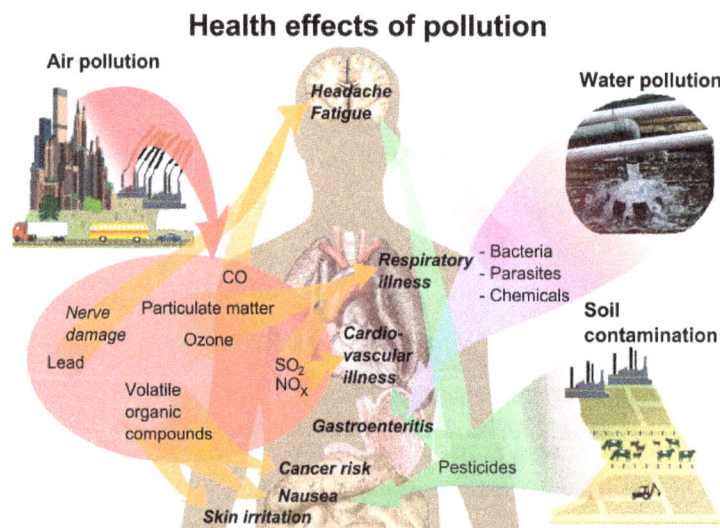

Pesticides are implicated in a range of impacts on human health due to pollution

The effects of pesticides on human health depend on the toxicity of the chemical and the length and magnitude of exposure. Farm workers and their families experience the greatest exposure to agricultural pesticides through direct contact. Every human contains pesticides in their fat cells.

Children are more susceptible and sensitive to pesticides, because they are still developing and have a weaker immune system than adults. Children may be more exposed due to their closer proximity to the ground and tendency to put unfamiliar objects in their mouth. Hand to mouth contact depends on the child's age, much like lead exposure. Children under the age of six months are more apt to experience exposure from breast milk and inhalation of small particles. Pesticides tracked into the home from family members increase the risk of exposure. Toxic residue in food may contribute to a child's exposure. The chemicals can bioaccumulate in the body over time.

Exposure effects can range from mild skin irritation to birth defects, tumors, genetic changes, blood and nerve disorders, endocrine disruption, coma or death. Developmental effects have been associated with pesticides. Recent increases in childhood cancers in throughout North America, such as leukemia, may be a result of somatic cell mutations. Insecticides targeted to disrupt insects can have harmful effects on mammalian nervous systems. Both chronic and acute alterations have been observed in exposees. DDT and its breakdown product DDE disturb estrogenic activity and possibly lead to breast cancer. Fetal DDT exposure reduces male penis size in animals and can produce undescended testicles. Pesticide can affect fetuses in early stages of development, in utero and even if a parent was exposed before conception. Reproductive disruption has the potential to occur by chemical reactivity and through structural changes.

Persistent Organic Pollutants

Persistent organic pollutants (POPs) are compounds that resist degradation and thus remain in the environment for years. Some pesticides, including aldrin, chlordane, DDT, dieldrin, endrin, heptachlor, hexachlorobenzene, mirex and toxaphene, are considered POPs. Some POPs have the ability to volatilize and travel great distances through the atmosphere to become deposited in remote regions. Such chemicals may have the ability to bioaccumulate and biomagnify and can bioconcentrate (i.e. become more concentrated) up to 70,000 times their original concentrations. POPs can affect non-target organisms in the environment and increase risk to humans by disruption in the endocrine, reproductive, and immune systems.

Pest Resistance

Pests may evolve to become resistant to pesticides. Many pests will initially be very susceptible to pesticides, but following mutations in their genetic makeup become resistant and survive to reproduce.

Resistance is commonly managed through pesticide rotation, which involves alternating among pesticide classes with different modes of action to delay the onset of or mitigate existing pest resistance.

Pest Rebound and Secondary Pest Outbreaks

Non-target organisms can also be impacted by pesticides. In some cases, a pest insect that is controlled by a beneficial predator or parasite can flourish should an insecticide application kill both pest and beneficial populations. A study comparing biological pest control and pyrethroid insecticide for diamondback moths, a major cabbage family insect pest, showed that the pest population rebounded due to loss of insect predators, whereas the biocontrol did not show the same effect. Likewise, pesticides sprayed to control mosquitoes may temporarily depress mosquito populations, however they may result in a larger population in the long run by damaging natural controls. This phenomenon, wherein the population of a pest species rebounds to equal or greater numbers than it had before pesticide use, is called pest resurgence and can be linked to elimination of its predators and other natural enemies.

Loss of predator species can also lead to a related phenomenon called secondary pest outbreaks, an increase in problems from species that were not originally a problem due to loss of their predators or parasites. An estimated third of the 300 most damaging insects in the US were originally secondary pests and only became a major problem after the use of pesticides. In both pest resurgence and secondary outbreaks, their natural enemies were more susceptible to the pesticides than the pests themselves, in some cases causing the pest population to be higher than it was before the use of pesticide.

Eliminating Pesticides

Many alternatives are available to reduce the effects pesticides have on the environment. Alternatives include manual removal, applying heat, covering weeds with plastic, placing traps and lures, removing pest breeding sites, maintaining healthy soils that breed healthy, more resistant plants, cropping native species that are naturally more resistant to native pests and supporting biocontrol agents such as birds and other pest predators. In the United States, conventional pesticide use peaked in 1979, and by 2007, had been reduced by 25 percent from the 1979 peak level, while US agricultural output increased by 43 percent over the same period.

Biological controls such as resistant plant varieties and the use of pheromones, have been successful and at times permanently resolve a pest problem. Integrated Pest Management (IPM) employs chemical use only when other alternatives are ineffective. IPM causes less harm to humans and the environment. The focus is broader than on a specific pest, considering a range of pest control alternatives. Biotechnology can also be an innovative way to control pests. Strains can be genetically modified (GM) to increase their resistance to pests. However the same techniques can be used to increase pesticide resistance and was employed by Monsanto to create glyphosate-resistant strains of major crops. In 2010, 70% of all the corn that was planted was resistant to glyphosate; 78% of cotton, and 93% of all soybeans.

Pesticide Drift

Pesticide drift refers to the unintentional diffusion of pesticides and the potential negative effects of pesticide application—including: off-target contamination due to spray drift as well as runoff from plants/soil. This can lead to damage in human health, environmental contamination, and property damage.

Types

With placement (localised) spraying of broad spectrum pesticides, wind drift must be minimized, and considerable efforts have been made recently to quantify and control spray drift from hydraulic nozzles. Conversely, wind drift is also an efficient mechanism for moving droplets of an appropriate size range to their targets over a wide area with ultra-low volume (ULV) spraying.

Himel (1974) made a distinction between exo-drift (the transfer of spray out of the target area) and endo-drift, where the active ingredient (AI) in droplets falls into the target area, but does not reach the biological target. Endo-drift is volumetrically more significant and may therefore cause greater ecological contamination (e.g. where chemical pesticides pollute ground water).

Herbicide Volatilisation

Herbicide volatilisation refers to evaporation or sublimation of a volatile herbicide. The effect of gaseous chemical is lost at its intended place of application and may move downwind and affect other plants not intended to be affected causing crop damage. Herbicides vary in their susceptibility to volatilisation. Prompt incorporation of the herbicide into the soil may reduce or prevent volatilisation. Wind, temperature, and humidity also affect the rate of volatilisation with humidity reducing in. 2,4-D and dicamba are commonly used chemicals that are known to be subject to volatilisation but there are many others. Application of herbicides later in the season to protect herbicide-resistant genetically modified plants increases the risk of volatilisation as the temperature is higher and incorporation into the soil impractical.

Public Concern

Although there has been much public concern and research into spray drift, several studies have concluded that point source pollution (e.g. pesticides entering bodies of water following spillage of concentrate or *rinsate*) can cause greatest environmental harm.

References

- Gosden Emily (29 March 2014) Bees and the crops they pollinate are at risk from climate change, IPCC report to warn The Daily Telegraph, Retrieved 30 March 2014

- Office of the Press Secretary (June 20, 2014). "The Economic Challenge Posed by Declining Pollinator Populations" (Factsheet). The White House. Retrieved 31 August 2015.

- Lawrence, Dune (February 13, 2007), Chinese develop taste for organic food: Higher cost no barrier to safer eating. Bloomberg News, International Herald Tribune Retrieved on 2007-10-25.

- McCauley LA, Anger WK, Keifer M, Langley R, Robson MG, Rohlman D (2006). "Studying health outcomes in farmworker populations exposed to pesticides". Environmental Health Perspectives. 114 (3): 953–960. doi:10.1289/ehp.8526. Retrieved 2007-09-15.

- Pulaski A (May 26, 2006), EPA workers blast agency's rulings on deadly pesticides: Letter sent to EPA administrator Stephen L. Johnson by unions representing 9,000 EPA scientists. The Oregonian, Mindfully.org Retrieved on 2007-10-10.

- Cornell University, College of Veterinary Medicine (March 1999), Consumer concerns about pesticides in food. Fact Sheet #24. Retrieved on 2007-10-25.

- Codex Alimentarius Commission Code of Ethics for International Trade in Food. CAC/RCP 20-1979 (Rev. 1-1985). Retrieved on 2007-10-25.

- U.S. Environmental Protection Agency (March 27, 2007), Pesticides and food: What the pesticide residue limits are on food. epa.gov. Retrieved on September 15, 2007.

- U.S. Environmental Protection Agency (July 24, 2007), Setting tolerances for pesticide residues in foods. epa.gov. Retrieved on September 15, 2007.

- Ramesh C. Gupta (28 April 2011). Toxicology of Organophosphate & Carbamate Compounds. Academic Press. pp. 352–353. ISBN 978-0-08-054310-9.

- Denis Hamilton; Stephen Crossley (14 May 2004). Pesticide Residues in Food and Drinking Water: Human Exposure and Risks. John Wiley & Sons. p. 280. ISBN 978-0-470-09160-9.

- Lewis A. Owen; Professor Kevin T Pickering; Kevin T. Pickering (1 March 2006). An Introduction to Global Environmental Issues. Routledge. p. 197. ISBN 978-1-134-76919-3.

8

Impacts of Pesticides Toxicity on Bees

Bees are natural agents responsible for pollination, which in turn allows for genetic variation. The use of pesticides has been linked to a decline in bee populations across the globe. This chapter studies the reasons behind the fall in bee colonies by elaborating on topics such as pesticide toxicity to bees, colony collapse disorder and pesticide formulation.

Pesticide Toxicity to Bees

Pesticides vary in their effects on bees. Contact pesticides are usually sprayed on plants and can kill bees when they crawl over sprayed surfaces of plants or other areas around it. Systemic pesticides, on the other hand, are usually incorporated into the soil or onto seeds and move up into the stem, leaves, nectar, and pollen of plants.

Of contact pesticides, dust and wettable powder pesticides tend to be more hazardous to bees than solutions or emulsifiable concentrates. When a bee comes in contact with pesticides while foraging, the bee may die immediately without returning to the hive. In this case, the queen bee, brood, and nurse bees are not contaminated and the colony survives. Alternatively, the bee may come into contact with an insecticide and transport it back to the colony in contaminated pollen or nectar or on its body, potentially causing widespread colony death.

Actual damage to bee populations is a function of toxicity and exposure of the compound, in combination with the mode of application. A systemic pesticide, which is incorporated into the soil or coated on seeds, may kill soil-dwelling insects, such as grubs or mole crickets as well as other insects, including bees, that are exposed to the leaves, fruits, pollen, and nectar of the treated plants.

Pesticides are linked to Colony Collapse Disorder and are now considered a main cause, and the toxic effects of Neonicotinoids on bees are confirmed. Currently, many studies are being conducted to further understand the toxic effects of pesticides on bees. Agencies such as the EPA and EFSA are making action plans to protect bee health in response to calls from scientists and the public to ban or limit the use of the pesticides with confirmed toxicity.

Classification

Insecticide toxicity is generally measured using acute contact toxicity values LD_{50} – the exposure level that causes 50% of the population exposed to die. Toxicity thresholds are generally set at

- highly toxic (acute LD50 < 2µg/bee)

- moderately toxic (acute LD50 2 - 10.99µg/bee)

- slightly toxic (acute LD50 11 - 100μg/bee)

- nontoxic (acute LD50 > 100μg/bee) to adult bees.

Pesticide Toxicity

Acute Toxicity

The acute toxicity of pesticides on bees, which could be by contact or ingestion, is usually quantified by LD_{50}. Acute toxicity of pesticides causes a range of effects on bees, which can include agitation, vomiting, wing paralysis, arching of the abdomen similar to sting reflex, and uncoordinated movement. Some pesticides, including Neonicotinoids, are more toxic to bees and cause acute symptoms with lower doses compared to older classes of insecticides. Acute toxicity may depend on the mode of exposure, for instance, many pesticides cause toxic effects by contact while Neonicotinoids are more toxic when consumed orally. The acute toxicity, although more lethal, is less common than sub-lethal toxicity or cumulative effects.

Sub-lethal Toxicity

Field exposure of bees to pesticides, especially with relation to neonicotinoids, is most commonly sub-lethal. Sub-lethal effects to honey bees are of major concern and include behavioral disruptions such as disorientation, reduced foraging, impaired memory and learning, and a shift in communication behaviors. Additional sub-lethal effects may include compromised immunity of bees and delayed development.

Cumulative and Chronic Effects

Neonicotinoids are especially likely to cause cumulative effects on bees due to their mechanism of function as this pesticide group works by binding to nicotinic acetylcholine receptors in the brains of the insects, and such receptors are particularly abundant in bees. Over-accumulation of acetylcholine results in paralysis and death.

Colony Collapse Disorder

Signs promote awareness of Colony Collapse Disorder and the importance of bees.

Colony collapse disorder (CCD) is a syndrome that is characterized by the sudden loss of adult bees from the hive. Many possible explanations for CCD have been proposed, but no one primary cause has been found. The US Department of Agriculture (USDA) has indicated in a report to Congress that a combination of factors may be causing CCD, including pesticides, pathogens, and parasites, all of which have been found at high levels in affected bee hives.

Colony Collapse Disorder has more implication than the extinction of some bee species; the disappearance of honeybees can cause catastrophic health and financial impacts. One mouthful in three of the food we eat may depend directly or indirectly on pollination by honeybee. Honeybee pollination has an estimated value of more than $14 billion annually to the United States agriculture. Honeybees are required for pollinating many crops, which range from nuts to vegetables and fruits, that are necessary for human and animal diet.

The EPA updated their guidance for assessing pesticide risks to honeybees in 2014. For the EPA, when certain pesticide use patterns or triggers are met, current test requirements include the honey bee acute contact toxicity test, the honey bee toxicity of residues on foliage test, and field testing for pollinators. EPA guidelines have not been developed for chronic or acute oral toxicity to adult or larval honey bees. On the other hand, the PMRA (Pest Management Regulatory Agency) requires both acute oral and contact honey bee adult toxicity studies when there is potential for exposure for insect pollinators. Primary measurement endpoint derived from the acute oral and acute contact toxicity studies is the median lethal dose for 50% of the organisms tested (i.e., LD_{50}), and if any biological effects and abnormal responses appear, including sub-lethal effects, other than the mortality, it should be reported.

The EPA's testing requirements do not account for sub-lethal effects to bees or effects on brood or larvae. Their testing requirements are also not designed to determine effects in bees from exposure to systemic pesticides. With colony collapse disorder, whole hive tests in the field are needed in order to determine the effects of a pesticide on bee colonies. To date, there are very few scientifically valid whole hive studies that can be used to determine the effects of pesticides on bee colonies because the interpretation of such whole-colony effects studies is very complex and relies on comprehensive considerations of whether adverse effects are likely to occur at the colony level.

A March 2012 study conducted in Europe, in which minuscule electronic localization devices were fixed on bees, has shown that, even with very low levels of pesticide in the bee's diet, a high proportion of bees (more than one third) suffers from orientation disorder and is unable to come back to the hive. The pesticide concentration was order of magnitudes smaller than the lethal dose used in the pesticide's current use. The pesticide under study, brand-named "Cruiser" in Europe (thiamethoxam, a neonicotinoid insecticide), although allowed in France by annually renewed exceptional authorization, could be banned in the coming years by the European Commission.

April 2013 the EU decided to restrict thiamethoxam, clothianidin, and imidacloprid.

Bee Kill Rate Per Hive

The kill rate of bees in a single bee hive can be classified as:

> < 100 bees per day - normal die off rate

200-400 bees per day - low kill

500-900 bees per day - moderate kill

> 1000 bees per day - high kill

Pesticides Formulations

Pesticides come in different formulations:

- Dusts (D)

- Wettable powders (WP)

- Soluble powders (SP)

- Emulsifiable concentrates (EC)

- Solutions (LS)

- Granulars (G)

Pesticides

Common name (ISO)	Examples of Brand names	Pesticide Class	length of residual toxicity	Comments	Bee toxicity
Sulfoxaflor		Sulfoximine			
Aldicarb	Temik	Carbamate		apply 4 weeks before bloom	Relatively nontoxic
Carbaryl	Sevin, (b) Sevin XLR	Carbamate	High risk to bees foraging even 10 hours after spraying; 3 – 7 days (b) 8 hours @ 1.5 lb/acre (1681 g/Ha) or less.	Bees poisoned with carbaryl can take 2–3 days to die, appearing inactive as if cold. Sevin should never be sprayed on flowering crops, especially if bees are active and the crop requires pollination. Less toxic formulations exist.	highly toxic
Carbofuran	Furadan	Carbamate	7 – 14 days	U.S. Environmental Protection Agency ban on use on crops grown for human consumption (2009) carbofuran (banned in granular form)	highly toxic
Methomyl	Lannate, Nudrin	Carbamate	2 hours	Should never be sprayed on flowering crops especially if bees are active and the crop requires pollination.	highly toxic

Common name (ISO)	Examples of Brand names	Pesticide Class	length of residual toxicity	Comments	Bee toxicity
Methiocarb	Mesurol	Carbamate			highly toxic
Mexacarbate	Zectran	Carbamate			highly toxic
Pirimicarb	Pirimor, Aphox	Carbamate			Relatively nontoxic
Propoxur	Baygon	Carbamate		Propoxur is highly toxic to honey bees. The LD50 for bees is greater than one ug/honey bee.	highly toxic
Acephate	Orthene	Organophosphate	3 days	Acephate is a broad-spectrum insecticide and is highly toxic to bees and other beneficial insects.	Moderately toxic
Azinphos-methyl	Guthion, Methyl-Guthion	Organophosphate	2.5 days	banned in the European Union since 2006.	highly toxic
Chlorpyrifos	Dursban, Lorsban	Organophosphate		banned in the US for home and garden use Should never be sprayed on flowering crops especially if bees are active and the crop requires pollination.	highly toxic
Coumaphos	Checkmite	Organophosphate		This is an insecticide that is used inside the beehive to combat varroa mites and small hive beetles, which are parasites of the honey bee. Overdoses can lead to bee poisoning.	Relatively nontoxic
Demeton	Systox	Organophosphate	<2 hours		highly toxic
Demeton-S-methyl	Meta-systox	Organophosphate			Moderately toxic
Diazinon	Spectracide	Organophosphate		Sale of diazinon for residential use was discontinued in the U.S. in 2004. Should never be sprayed on flowering crops especially if bees are active and the crop requires pollination.	highly toxic
Dicrotophos	Bidrin	Organophosphate		Dicrotophos toxicity duration is about one week.	highly toxic
Dichlorvos	DDVP, Vapona	Organophosphate			highly toxic

Common name (ISO)	Examples of Brand names	Pesticide Class	length of residual toxicity	Comments	Bee toxicity
Dimethoate	Cygon, De-Fend	Organophosphate	3 days	Should never be sprayed on flowering crops especially if bees are active and the crop requires pollination.	highly toxic
Fenthion	Entex, Baytex, Baycid, Dalf, DMPT, Mercaptophos, Prentox, Fenthion 4E, Queletox,Lebaycid	Organophosphate		Should never be sprayed on flowering crops especially if bees are active and the crop requires pollination.	highly toxic
Fenitrothion	Sumithion	Organophosphate			highly toxic
Fensulfothion	Dasanit	Organophosphate			highly toxic
Fonofos	Dyfonate EC	Organophosphate	3 hours	List of Schedule 2 substances (CWC)	highly toxic
Malathion	Malathion USB, ~ EC, Cythion, maldison, mercaptothion	Organophosphate	>8 fl oz/acre (58 L/km²) □ 5.5 days	Malathion is highly toxic to bees and other beneficial insects, some fish, and other aquatic life. Malathion is moderately toxic to other fish and birds, and is considered low in toxicity to mammals.	highly toxic
Methamidophos	Monitor, Tameron	Organophosphate		Should never be sprayed on flowering crops especially if bees are active and the crop requires pollination.	highly toxic
Methidathion	Supracide	Organophosphate		Should never be sprayed on flowering crops especially if bees are active and the crop requires pollination.	highly toxic
Methyl parathion	Parathion, Penncap-M	Organophosphate	5–8 days	It is classified as a UNEP Persistent Organic Pollutant and WHO Toxicity Class, "Ia, Extremely Hazardous".	highly toxic
Mevinphos	Phosdrin	Organophosphate			highly toxic
Monocrotophos	Azodrin	Organophosphate		Should never be sprayed on flowering crops especially if bees are active and the crop requires pollination.	highly toxic
Naled	Dibrom	Organophosphate	16 hours		highly toxic

Common name (ISO)	Examples of Brand names	Pesticide Class	length of residual toxicity	Comments	Bee toxicity
Omethoate		Organophosphate		Should never be sprayed on flowering crops especially if bees are active and the crop requires pollination.	highly toxic
Oxydemeton-methyl	Metasystox-R	Organophosphate	<2 hours		highly toxic
Phorate	Thimet EC	Organophosphate	5 hours		highly toxic
Phosmet	Imidan	Organophosphate		Phosmet is very toxic to honeybees.	highly toxic
Phosphamidon	Dimecron	Organophosphate			highly toxic
Pyrazophos	Afugan	Organophosphate	fungicide		highly toxic
Tetrachlorvinphos	Rabon, Stirofos, Gardona, Gardcide	Organophosphate			highly toxic
Trichlorfon, Metrifonate	Dylox, Dipterex	Organophosphate		3 – 6 hours	Relatively nontoxic
Permethrin	Ambush, Pounce	Synthetic pyrethroid	1 – 2 days	safened by repellency under arid conditions. Permethrin is also the active ingredient in insecticides used against the Small hive beetle, which is a parasite of the beehive in the temperate climate regions.	highly toxic
Cypermethrin	Ammo, Raid	Synthetic pyrethroid	Less than 2 hours	Cypermethrin is found in many household ant and cockroach killers, including Raid and ant chalk.	highly toxic
Fenvalerate	Asana, Pydrin	Synthetic pyrethroid	1 day	safened by repellency under arid conditions	highly toxic
Resmethrin	Chrysron, Crossfire, Pynosect, Raid Flying Insect Killer, Scourge, Sun-Bugger #4, SPB-1382, Synthrin, Syntox, Vectrin, Whitmire PT-110	Synthetic pyrethroid		Resmethrin is highly toxic to bees, with an LD50 of 0.063 ug/bee.	highly toxic
Methoxychlor	DMDT, Marlate	Chlorinated cyclodiene	2 hours	available as a General Use Pesticide	highly toxic
Endosulfan	Thiodan	Chlorinated cyclodiene	8 hours	banned in European Union (2007?), New Zealand (2009)	moderately toxic

Common name (ISO)	Examples of Brand names	Pesticide Class	length of residual toxicity	Comments	Bee toxicity
Clothianidin	Poncho	Neonicotinoid		Banned in Germany In June 2008, the Federal Ministry of Food, Agriculture and Consumer Protection (Germany) suspended the registration of eight neonicotinoid pesticide seed treatment products used in oilseed rape and sweetcorn, a few weeks after honey bee keepers in the southern state of Baden Württemberg reported a wave of honey bee deaths linked to one of the pesticides, clothianidin.	Highly Toxic
Thiamethoxam	Actara	Neonicotinoid		Clothianidin is a major metabolite of Thiamethoxam. A two-year study published in 2012 showed the presence of clothianidin and thiamethoxam in bees found dead in and around hives situated near agricultural fields. Other bees at the hives exhibited tremors and uncoordinated movement and convulsions, all signs of insecticide poisoning.	Highly Toxic
Imidacloprid	Confidor, Gaucho, Kohinor, Admire, Advantage, Merit, Confidor, Hachikusan, Amigo, SeedPlus (Chemtura Corp.), Monceren GT, Premise, Prothor, and Winner	Neonicotinoid		Banned in France since 1999	highly toxic
Dicofol		Acaricide			Relatively nontoxic
Petroleum oils					Relatively nontoxic
2,4-D	ingredient in over 1,500 products	Synthetic auxin herbicide			Relatively nontoxic

Neonicotinoids

Neonicotinoids are one of the leading suspected causes of colony collapse disorder in honey bees. The specific causes are unclear, however, there has been some research to show that neonicotinoids have deleterious health effects on colony queens. Managed honeybee colonies are colonies that are "man-made." That is, they are not naturally occurring. Rather they are raised and rented out to farmers.

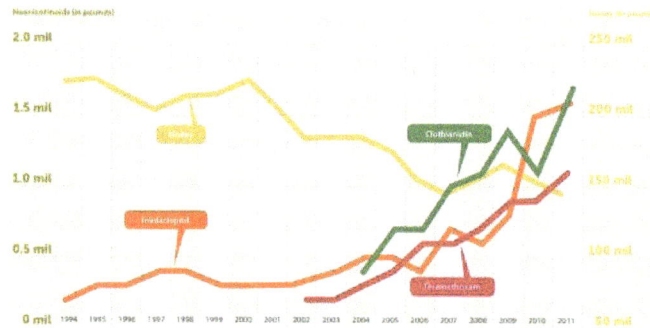

Decline of annual honey production between 2000 and 2011 by one third from 221 million to 148 million pounds with increased Neonicotinoids use in the United States.

There is some controversy surrounding the specific issue of whether or not neonicotinoids actually do negatively affect managed honeybee colonies. Perhaps one of the most popular studies showing a significant association between colony collapse disorder and neonicotinoids is 'Sub-lethal exposure to neonicotinoids impaired honeybees Winterization before proceeding to colony collapse disorder' by Chensheng Lu. Chengsheng became somewhat of a folk hero among environmental activists after his Harvard study was published, however, there has been some dissent from bee researchers in Australia, Canada, and even the USDA. That is not to say that Mr. Lu's findings are not accurate. The problem is that there simply is not a consensus yet on the real association between neonicotinoids and colony collapse disorder.

Common Insecticides Toxic to Bees and Used on Soybeans

Many insecticides used against soybean aphids are highly toxic to bees.

- Orthene 75S (Acephate)

- Address 75 WSP (Acephate)

- Sevin (Carbaryl)

- Lorsban 4E, Chlorpyrifos, Eraser, Govern, Nufos, Pilot, Warhawk, Whirlwind and Yuma (Chlorpyrifos)

- Dimate (Dimethoate)

- Steward 1.25 SC (Indoxacarb)

- Lannate (Methomyl)

- Cheminova Methyl 4EC (Methyl Parathion)

- Penncap M (microencapsulated Methyl Parathion)

- Tracer (Spinosad)

- Tombstone (Cyfluthrin)

- Baythroid XL (Beta-cyfluthrin)

- Delta Gold (Deltamethrin)

Highly Toxic and Banned in The Us

- Aldrin banned by US EPA in 1974

- Dieldrin banned by US EPA in 1974

- Heptachlor

- Lindane, BHC (banned in California). Lindane was also denied re-registration for agricultural use in the US by the EPA in 2006

Epa Proposal to Protect Bees from Acutely Toxic Pesticides in The US

The EPA is proposing to prohibit the application of certain pesticides and herbicides known toxic to bees during pollination periods when crops are in bloom. Growers routinely contract with honeybee keepers to bring in bees to pollinate their crops that require insect pollination. Bees are typically present during the period the crops are in bloom. Application of pesticides during this period can significantly affect the health of bees. These restrictions are expected to reduce the likelihood of high levels of pesticide exposure and mortality for bees providing pollination services. Moreover, the EPA believes these additional measures to protect bees providing pollination services will protect other pollinators as well.

The proposed restrictions would apply to all products that have liquid or dust formulations as applied, foliar use (applying pesticides directly to crop leaves) directions for use on crops, and active ingredients that have been determined via testing to have high toxicity for bees (less than 11 micrograms per bee). These restrictions would not replace already existing more restrictive, chemical-specific, and bee-protective provisions. Additionally, the proposed label restrictions would not apply to applications made in support of a government-declared public health response, such as use for wide area mosquito control. There would be no other exceptions to these proposed restrictions.

General Measures to Prevent Pesticides Bee Kills

Application of Pesticides at Evening or Night

Avoiding pesticide application directly to blooming flowers as much as possible can help limit the exposure of honeybees to toxic materials as honeybees are attracted to all types of blooming flowers. If blooming flowers must be sprayed with pesticides for any reason, they should be sprayed in the evening or night hours as bees are not in the field at that time. Usual foraging hours of honeybees are when the temperature is above 55-60 °F during the daytime, and by the evening, the bees return to the hives.

Colony Collapse Disorder

Colony collapse disorder (CCD) is the phenomenon that occurs when the majority of worker bees in a colony disappear and leave behind a queen, plenty of food and a few nurse bees to care for the remaining immature bees and the queen. While such disappearances have occurred throughout the history of apiculture, and were known by various names (disappearing disease, spring dwindle, May disease, autumn collapse, and fall dwindle disease), the syndrome was renamed colony collapse disorder in late 2006 in conjunction with a drastic rise in the number of disappearances of western honey bee (*Apis mellifera*) colonies in North America. European beekeepers observed similar phenomena in Belgium, France, the Netherlands, Greece, Italy, Portugal, and Spain, Switzerland and Germany, albeit to a lesser degree, and the Northern Ireland Assembly received reports of a decline greater than 50%.

Colony collapse disorder causes significant economic losses because many agricultural crops (although no staple foods) worldwide are pollinated by western honey bees. According to the Agriculture and Consumer Protection Department of the Food and Agriculture Organization of the United Nations, the worth of global crops with honey bee's pollination was estimated to be close to $200 billion in 2005. Shortages of bees in the US have increased the cost to farmers renting them for pollination services by up to 20%.

In the six years leading up to 2013, more than 10 million beehives were lost, often to CCD, nearly twice the normal rate of loss.

Several possible causes for CCD have been proposed, but no single proposal has gained widespread acceptance among the scientific community. Suggested causes include: infections with *Varroa* and *Acarapis* mites; malnutrition; various pathogens; genetic factors; immunodeficiencies; loss of habitat; changing beekeeping practices; or a combination of factors. A large amount of speculation has surrounded a family of pesticides called neonicotinoids as having caused CCD.

History

Honey bee on camas flower.

Limited occurrences resembling CCD have been documented as early as 1869 and this set of symptoms has, in the past several decades, been given many different names (disappearing disease, spring dwindle, May disease, autumn collapse, and fall dwindle disease). Most recently, a similar phenomenon in the winter of 2004/2005 occurred, and was attributed to varroa mites (the "vampire mite" scare), though this was never ultimately confirmed. The cause of the appearance of this syndrome has never been determined. Upon recognition that the syndrome does not seem to be seasonally restricted, and that it may not be a "disease" in the standard sense—that there may not be a specific causative agent—the syndrome was renamed.

A well-documented outbreak of colony losses spread from the Isle of Wight to the rest of the UK in 1906. These losses later were attributed to a combination of factors, including adverse weather, intensive apiculture leading to inadequate forage, and a new infection, the chronic bee paralysis virus, but at the time, the cause of this agricultural beekeeping problem was similarly mysterious and unknown.

Reports show this behavior in hives in the US in 1918 and 1919. Coined "mystery disease" by some, it eventually became more widely known as "disappearing disease". Oertel, in 1965, reported that hives afflicted with disappearing disease in Louisiana had plenty of honey in the combs, although few or no bees were present, discrediting reports that attributed the disappearances to lack of food.

From 1972 to 2006, dramatic reductions continued in the number of feral honey bees in the U.S. and a significant though somewhat gradual decline in the number of colonies maintained by beekeepers. This decline includes the cumulative losses from all factors, such as urbanization, pesticide use, tracheal and *Varroa* mites, and commercial beekeepers' retiring and going out of business. However, in late 2006 and early 2007, the rate of attrition was alleged to have reached new proportions, and the term "colony collapse disorder" began to be used to describe this sudden rash of disappearances (sometimes referred to as "spontaneous hive collapse" or the "Mary Celeste syndrome" in the United Kingdom).

Losses had remained stable since the 1990s at 17%–20% per year attributable to a variety of factors, such as mites, diseases, and management stress. The first report of CCD was in mid-November 2006 by a Pennsylvania beekeeper overwintering in Florida. By February 2007, large commercial migratory beekeepers in several states had reported heavy losses associated with CCD. Their reports of losses varied widely, ranging from 30% to 90% of their bee colonies; in some cases, beekeepers reported losses of nearly all of their colonies with surviving colonies so weakened that they might no longer be able to pollinate or produce honey.

Losses were reported in migratory operations wintering in California, Florida, Oklahoma, and Texas. In late February, some larger nonmigratory beekeepers in the mid-Atlantic and Pacific Northwest regions also reported significant losses of more than 50%. Colony losses also were reported in five Canadian provinces, several European countries, and countries in South and Central America and Asia. In 2010, the USDA reported that data on overall honey bee losses for 2010 indicated an estimated 34% loss, which is statistically similar to losses reported in 2007, 2008, and 2009. Fewer colony losses occurred in the U.S. over the winter of 2013-2014 than in recent years. Total losses of managed honey bee colonies from all causes were 23.2% nationwide, a marked improvement over the 30.5% loss reported for the winter of 2012-2013 and the eight-year average loss of 29.6%.

After bee populations dropped 23% in the winter of 2013, the Environmental Protection Agency and Department of Agriculture formed a task force to address the issue. In the six years leading up to 2013, more than 10 million beehives were lost, often to CCD, nearly twice the normal rate of loss. However, according to Syngenta, the total number of beehives worldwide continues to grow. An insecticide produced by Syngenta was banned by the European Commission in 2013 for use in crops pollinated by bees. Syngenta together with Bayer is challenging this ban in court.

Signs and Symptoms

In collapsed colonies, CCD is suspected when a complete absence of adult bees is found in colonies, with little or no buildup of dead bees in the hive or in front of the hive. A colony which has collapsed from CCD is generally characterized by all of these conditions occurring simultaneously:

Visit to a bee colony in West Virginia

- Presence of capped brood in abandoned colonies: Bees normally will not abandon a hive until the capped brood have all hatched.

- Presence of food stores, both honey and bee pollen:

  o which are not immediately robbed by other bees

  o which when attacked by hive pests such as wax moth and small hive beetle, the attack is noticeably delayed

- Presence of the queen bee: If the queen is not present, the hive died because it was queenless, which is not considered CCD.

Precursor symptoms that may arise before the final colony collapse are:

- Insufficient workforce to maintain the brood that is present

- Workforce seems to be made up of young adult bees

- The colony members are reluctant to consume provided feed, such as sugar syrup and protein supplement.

Scope and Distribution

North America

The National Agriculture Statistics Service reported 2.44 million honey-producing hives were in the United States in February 2008, down from 4.5 million in 1980, and 5.9 million in 1947, though these numbers underestimate the total number of managed hives, as they exclude several thousand hives managed for pollination contracts only, and also do not include hives managed by beekeepers owning fewer than five hives. This under-representation may be offset by the practice of counting some hives more than once; hives that are moved to different states to produce honey are counted in each state's total and summed in total counts.

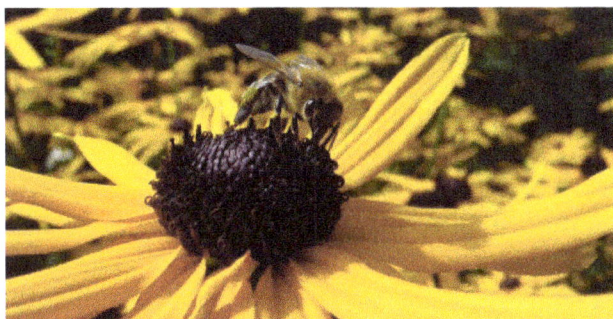

Bee (Apoidea) on flower (Rudbeckia spec.)

Non-CCD winter losses as high as 50% have occurred in some years and regions (e.g., 2000–2001 in Pennsylvania). Normal winter losses are typically considered to be in the range of 15–25%. In many cases, beekeepers reporting significant losses of bees did not experience true CCD, but losses due to other causes.

In 2007 in the US, at least 24 different states had reported at least one case of CCD. In a 2007 survey of 384 responding beekeepers from 13 states, 23.8% met the specified criterion for CCD (that 50% or more of their dead colonies were found without bees and/or with very few dead bees in the hive or apiary).

In the US in 2006–2007, CCD-suffering operations had a total loss of 45% compared to the total loss of 25% of all colonies experienced by non-CCD suffering beekeepers.

A 2007–2008 survey of over 19% of all colonies revealed a total loss of 35.8%. Operations that pollinated almonds lost, on average, the same number of colonies as those that did not. The 37.9% of operations that reported having at least some of their colonies die with a complete lack of bees had a total loss of 40.8% of colonies compared to the 17.1% loss reported by beekeepers without this symptom. Large operations were more likely to have this symptom, suggesting a contagious condition may be a causal factor. About 60% of all colonies that were reported dead in this survey died without the presence of dead bees in the hive, thus possibly suffered from CCD.

In 2010, the USDA reported that data on overall honey bee losses for the year indicate an estimated 34% loss, which is statistically similar to losses reported in 2007, 2008, and 2009. In 2011, the loss was 30%. In 2012–2013, CCD was blamed for the loss of about half of the US honey bee hives, far more than the 33% losses observed on average over previous years.

During the spring of 2015, President Barack Obama unveiled the very first national strategy for improving the health of bees and other key pollinators. The plan calls for restoring 7 million acres of bee habitat with a variety of different plants for bees to eat. The administration is also proposing spending $82.5 million for honey bee research.

Europe

According to the European Food Safety Authority (EFSA), in 2007, the United Kingdom had 274,000 hives, Italy had 1,091,630, and France 1,283,810. In 2008, the British Beekeepers Association reported the bee population in the United Kingdom dropped by around 30% between 2007 and 2008, and an EFSA study revealed that in Italy the mortality rate was 40–50%. However, EFSA officials point out the figures are not very reliable because before the bees started dying, no harmonisation was used in the way different countries collected statistics on their bee populations. At that time (2008), the reports blamed the high death rate on the varroa mite, two seasons of unusually wet European summers, and some pesticides.

In 2009, Tim Lovett, president of the British Beekeepers' Association, said: "Anecdotally, it is hugely variable. There are reports of some beekeepers losing almost a third of their hives and others losing none." John Chapple, chairman of the London Beekeepers' Association, put losses among his 150 members at between a fifth and a quarter. "There are still a lot of mysterious disappearances; we are no nearer to knowing what is causing them." The government's National Bee Unit continued to deny the existence of CCD in Britain; it attributes the heavy losses to the varroa mite and rainy summers that stop bees foraging for food.

An atlas of commercial geography (1913) (14781413295)

In 2010, David Aston of the British Beekeepers' Association stated, "We still do not believe CCD (which is now better defined) is a cause of colony losses in the UK, however we are continuing to experience colony losses, many if not most of which can be explained". He feels recent studies suggest "further evidence to the evolving picture that there are complex interactions taking place between a number of factors, pathogens, environmental, beekeeping practices and other stressors, which are causing honey bee losses described as CCD in the US".

Beekeepers in Scotland also reported losses for the past three years. Andrew Scarlett, a Perth-shire-based bee farmer and honey packer, lost 80% of his 1,200 hives during the 2009 winter. He attributed the losses to a virulent bacterial infection that quickly spread because of a lack of bee inspectors, coupled with sustained poor weather that prevented honey bees from building up sufficient pollen and nectar stores.

In Germany, where some of the first reports of CCD in Europe appeared, and where, according to the German national association of beekeepers, 40% of the honey bee colonies died, there was no scientific confirmation; in early May 2007, the German media reported no confirmed CCD cases seemed to have occurred in Germany.

At the end of May 2012, the Swiss government reported about half of the bee population had not survived the winter. The main cause of the decline was thought to be the parasite *Varroa destructor*.

Possible Causes

Honeybees in a hive. The hive was built under the fifth floor sun shade of a multystorey building. The photo was taken during daytime. Location- Kochi, Kerala, India

The mechanisms of CCD are still unknown, but many causes are currently being considered, such as pesticides, mites, fungi, beekeeping practices (such as the use of antibiotics or long-distance transportation of beehives), malnutrition, poor quality queens, starvation, other pathogens, and immunodeficiencies. The current scientific consensus is that no single factor is causing CCD, but that some of these factors in combination may lead to CCD either additively or synergistically.

In 2006, the Colony Collapse Disorder Working Group, based primarily at Pennsylvania State University, was established. Their preliminary report pointed out some patterns, but drew no strong conclusions. A survey of beekeepers early in 2007 indicated most hobbyist beekeepers believed that starvation was the leading cause of death in their colonies, while commercial beekeepers overwhelmingly believed invertebrate pests (*Varroa* mites, honey bee tracheal mites, and/or small hive beetles) were the leading cause of colony mortality. A scholarly review in June 2007 similarly addressed numerous theories and possible contributing factor, but left the issue unresolved.

In July 2007, the United States Department of Agriculture (USDA) released its "CCD Action Plan", which outlined a strategy for addressing CCD consisting of four main components: survey and data collection; analysis of samples; hypothesis-driven research; mitigation and preventive action. The first annual report of the U.S. Colony Collapse Disorder Steering Committee was published in 2009. It suggested CCD may be caused by the interaction of many agents in combination. The same year, the CCD Working Group published a comprehensive descriptive study that concluded: "Of the 61 variables quantified (including adult bee physiology, pathogen loads, and pesticide levels), no single factor was found with enough consistency to suggest one causal agent. Bees in CCD colonies had higher pathogen loads and were co-infected with more pathogens than control populations, suggesting either greater pathogen exposure or reduced defenses in CCD bees."

The second annual Steering Committee report was released in November 2010. The group reported, although many associations, including pesticides, parasites, and pathogens have been identified throughout the course of research, "it is becoming increasingly clear that no single factor alone is responsible for [CCD]". Their findings indicated an absence of damaging levels of the parasite *Nosema* or parasitic *Varroa* mites at the time of collapse. They did find an association of sublethal effects of some pesticides with CCD, including two common miticides in particular, coumaphos and fluvalinate, which are pesticides registered for use by beekeepers to control varroa mites. Studies also identified sublethal effects of neonicotinoids and fungicides, pesticides that may impair the bees' immune systems and may leave them more susceptible to bee viruses.

A 2015 review examined 170 studies on colony collapse disorder and stressors for bees, including pathogens, agrochemicals, declining biodiversity, climate change and more. The review concluded that "a strong argument can be made that it is the interaction among parasites, pesticides, and diet that lies at the heart of current bee health problems." Furthermore:

"Bees of all species are likely to encounter multiple stressors during their lives, and each is likely to reduce the ability of bees to cope with the others. A bee or bee colony that appears to have succumbed to a pathogen may not have died if it had not also been exposed to a sublethal dose of a pesticide and/or been subject to food stress (which might in turn be due to drought or heavy rain induced by climate change, or competition from a high density of honey bee hives placed nearby). Unfortunately, conducting well-replicated studies of the effects of multiple interacting stressors on bee colonies is exceedingly difficult. The number of stressor combinations rapidly becomes large, and exposure to stressors is hard or impossible to control with free-flying bees. Nonetheless, a strong argument can be made that it is the interaction among parasites, pesticides, and diet that lies at the heart of current bee health problems."

Pesticides

According to the USDA, pesticides may be contributing to CCD. A 2013 peer-reviewed literature review concluded neonicotinoids in the amounts typically used harm bees and safer alternatives are urgently needed. At the same time, other sources suggest the evidence is not conclusive, and that clarity regarding the facts is hampered by the role played by various issue advocates and lobby groups.

New Holland TL 90 with a field sprayer on a Narcissus field in Europe.

Scientists have long been concerned that pesticides, including possibly some fungicides, may have sublethal effects on bees, not killing them outright, but instead impairing their development and behavior. Of special interest is the class of insecticides called neonicotinoids, which contain the active ingredient imidacloprid, and other similar chemicals, such as clothianidin and thiamethoxam. Honey bees may be affected by such chemicals when they are used as a seed treatment because they are known to work their way through the plant up into the flowers and leave residues in the nectar. The doses taken up by bees are not lethal, but possible chronic problems could be caused by long-term exposure. Most corn grown in the US is treated with neonicotinoids, and a 2012 study found high levels of clothianidin in pneumatic planter exhaust. In the study, the insecticide was present in the soil of unplanted fields near those planted with corn and on dandelions growing near those fields. Another 2012 study also found clothianidin and imidacloprid in the exhaust of pneumatic seeding equipment.

A 2010 survey reported 98 pesticides and metabolites detected in aggregate concentrations up to 214 ppm in bee pollen; this figure represents over half of the individual pesticide incidences ever reported for apiaries. It was suggested that "while exposure to many of these neurotoxicants elicits acute and sublethal reductions in honey bee fitness, the effects of these materials in combinations and their direct association with CCD or declining bee health remains to be determined."

Evaluating pesticide contributions to CCD is particularly difficult for several reasons. First, the variety of pesticides in use in the different areas reporting CCD makes it difficult to test for all possible pesticides simultaneously. Second, many commercial beekeeping operations are mobile, transporting hives over large geographic distances over the course of a season, potentially exposing the colonies to different pesticides at each location. Third, the bees themselves place pollen and honey into long-term storage, effectively meaning a delay may occur from days to months before contaminated provisions are fed to the colony, negating any attempts to associate the appearance of symptoms with the actual time at which exposure to pesticides occurred.

Pesticides used on bee forage are far more likely to enter the colony by the pollen stores rather than nectar (because pollen is carried externally on the bees, while nectar is carried internally, and may kill the bee if too toxic), though not all potentially lethal chemicals, either natural or man-made, affect the adult bees; many primarily affect the brood, but brood die-off does not appear to be happening in CCD. Most significantly, brood are not fed honey, and adult bees consume relatively little pollen; accordingly, the pattern in CCD suggests, if contaminants or toxins from the environment 'are' responsible, it is most likely to be via the honey, as the adults are dying (or leaving), not the brood (though possibly effects of contaminated pollen consumed by juveniles may only show after they have developed into adults).

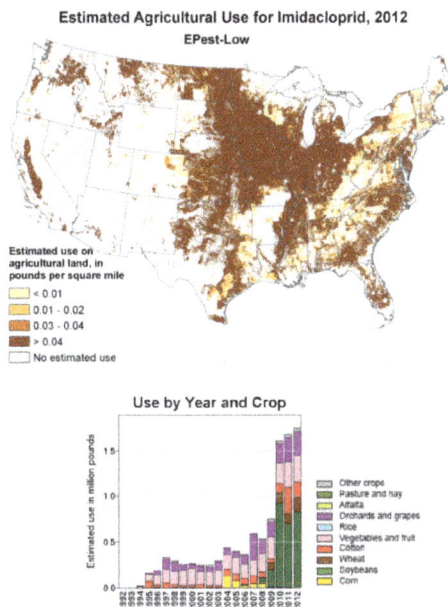

Imidacloprid map of use, USA, 2012 (estimated)

To date, most of the evaluation of possible roles of pesticides in CCD have relied on the use of surveys submitted by beekeepers, but direct testing of samples from affected colonies seems likely to be needed, especially given the possible role of systemic insecticides such as the neonicotinoid imidacloprid (which are applied to the soil and taken up into the plant's tissues, including pollen and nectar), which may be applied to a crop when the beekeeper is not present. The known effects of imidacloprid on insects, including honey bees, are consistent with the symptoms of CCD; for example, the effects of imidacloprid on termites include apparent failure of the immune system, and disorientation.

In Europe, the interaction of the phenomenon of "dying bees" with imidacloprid has been discussed for quite some time. A study from the "Comité Scientifique et Technique (CST)" was at the center of discussion, and led to a partial ban of imidacloprid in France. The imidacloprid pesticide Gaucho was banned in 1999 by the French Minister of Agriculture Jean Glavany, primarily due to concern over potential effects on honey bees. Subsequently, when fipronil, a phenylpyrazole insecticide and in Europe mainly labeled "Regent", was used as a replacement, it was also found to be toxic to bees, and banned partially in France in 2004.

In February 2007, about 40 French deputies, led by Jacques Remiller of the UMP, requested the creation of a parliamentary investigation commission on overmortality of bees, underlining that honey production had decreased by 1,000 tons a year for a decade. By August 2007, no investigation had opened. Five other insecticides based on fipronil were also accused of killing bees. However, the scientific committees of the European Union are still of the opinion "that the available monitoring studies were mainly performed in France and EU-member-states should consider the relevance of these studies for the circumstances in their country."

Around the same time French beekeepers succeeded in banning neonicotinoids, the Clinton administration permitted pesticides which were previously banned, including imidacloprid. In 2004, the Bush administration reduced regulations further and pesticide applications increased.

In 2005, a team of scientists led by the National Institute of Beekeeping in Bologna, Italy, found pollen obtained from seeds dressed with imidacloprid contain significant levels of the insecticide, and suggested the polluted pollen might cause honey bee colony death. Analysis of maize and sunflower crops originating from seeds dressed with imidacloprid suggest large amounts of the insecticide will be carried back to honey bee colonies. Sublethal doses of imidacloprid in sucrose solution have also been documented to affect homing and foraging activity of honey bees. Imidacloprid in sucrose solution fed to bees in the laboratory impaired their communication for a few hours. Sublethal doses of imidacloprid in laboratory and field experiment decreased flight activity and olfactory discrimination, and olfactory learning performance was impaired.

Research, in 2008, by scientists from Pennsylvania State University found high levels of the pesticides fluvalinate and coumaphos in samples of wax from hives, as well as lower levels of 70 other pesticides. These chemicals have been used to try to eradicate varroa mites, a bee pest that itself has been thought to be a cause of CCD. Researchers from Washington State University, under entomology professor Steve Sheppard in 2009, confirmed high levels of pesticide residue in hive wax and found an association between it and significantly reduced bee longevity.

The WSU work also focused on the impact of the microsporidian pathogen *Nosema ceranae*, the build-up of which was high in the majority of the bees tested, even after large doses of the antibiotic fumagillin. Penn State's Dr. Maryann Frazier said, "Pesticides alone have not shown they are the cause of CCD. We believe that it is a combination of a variety of factors, possibly including mites, viruses and pesticides."

In 2010, fipronil was blamed for the spread of CCD among bees, in a study by the Minutes-Association for Technical Coordination Fund in France, which found that even at very low nonlethal doses, this pesticide still impairs the ability to locate the hive, resulting in large numbers of foragers lost with every pollen-finding expedition, though no mention was made regarding any of the other symptoms of CCD; other studies, however, have shown no acute effect of fipronil on honey bees. Fipronil is designed to eliminate insects similar to bees, such as yellowjackets (*Vespula germanica*) and many other colonial pests by a process of 'toxic baiting', whereby one insect returning to the hive spreads the pesticide among the brood.

Honeycomb of honey bees with eggs and larvae. The walls of the cells have been removed. The larvae (drones) are about 3 or 4 days old.

A large 2010 survey of healthy and CCD-affected colonies also revealed elevated levels of pesticides in wax and pollen, but the amounts of pesticides were similar in both failing and healthy hives. They also confirmed suspected links between CCD and poor colony health, inadequate diet, and long-distance transportation. Studies continue to show very high levels of pathogens in CCD-affected samples and lower pathogen levels in unaffected samples, consistent with the empirical observation that healthy honey bee colonies normally fend off pathogens. These observations have led to the hypothesis that bee declines are resulting from immune suppression.

In 2010, a sequencing of the honey bee genome provided a possible explanation for the sensitivity of bees to pesticides. Its genome is deficient in the number of genes encoding detoxification enzymes, including cytochrome P450 monooxygenases (P450s), glutathione-S-transferases, and carboxylesterases.

In 2012, researchers announced findings that sublethal exposure to imidacloprid rendered honey bees significantly more susceptible to infection by the fungus *Nosema*, thereby suggesting a potential link to CCD, given that *Nosema* is increasingly considered to contribute to CCD.

Neonicotinoids may interfere with bees' natural homing abilities, causing them to become disoriented and preventing them from finding their way back to the hive.

Also, in 2012, researchers in Italy published findings that the pneumatic drilling machines that plant corn seeds coated with clothianidin and imidacloprid release large amounts of the pesticide into the air, causing significant mortality in foraging honey bees. According to the study, "Experimental results show that the environmental release of particles containing neonicotinoids can produce high exposure levels for bees, with lethal effects compatible with colony losses phenomena observed by beekeepers." Commonly used pesticides, such as the imidacloprid, reduce colony growth and new queen production in experimental exposure matched to field levels. Lu *et al.* (2012) reported they were able to replicate CCD with imidacloprid. Another neonicotinoid, thiamethoxam, causes navigational homing failure of foraging bees, with high mortality.

A 2012 *in situ* study provided strong evidence that exposure to sublethal levels of imidacloprid in high fructose corn syrup (HFCS) used to feed honey bees when forage is not available causes bees to exhibit symptoms consistent to CCD 23 weeks after imidacloprid dosing. The researchers suggested, "the observed delayed mortality in honey bees caused by imidacloprid in HFCS is a novel and plausible mechanism for CCD, and should be validated in future studies."

In March 2013, two studies were published showing that neonicotinoids affect bee long-term and short-term memory, suggesting a cause of action resulting in failure to return to the hive. In another study done in 2013, scientists reported that experiments suggested that exposure to the neonicotinoid pesticides clothianidin and imidicloprid results in increased levels of a particular protein in bees that inhibits a key molecule involved in the immune response, making the insects more susceptible to attack by harmful viruses. Growth in the use of neonicotinoid pesticides has roughly tracked rising bee deaths. In 2015, an 11-year British study showed a definitive relationship between increasing agricultural use of neonicotinoid and escalating honey bee colony losses at a landscape level. This is the first field study to establish a link between neonicotinoids and CCD.

Bees Collecting Pollen 2004-08-14

In July 2013, scientists from the University of Maryland and the US Department of Agriculture found that a combination of pesticides has been contaminating the pollen bees use to feed their hives. When researchers collected pollen from hives on the east coast, they discovered that it was contaminated (on average) with 9 different fungicides and pesticides, although scientists found a blend of 21 different agricultural chemicals in one sample of pollen. Eight ag chemicals were identified to be associated with increased risk of infection by *Nosema ceranae*.

A meta-analysis study published in February 2016 strongly suggests a pattern linking imidacloprid to sublethal effects on honey bees, stating: "trace dietary imidacloprid at field-realistic levels in nectar will have no lethal effects, but will reduce expected performance in honey bees by between 6 and 20%. Statistical power analysis showed that published field trials that have reported no effects on honey bees from neonicotinoids were incapable of detecting these predicted sublethal effects with conventionally accepted levels of certainty."

European Food Safety Authority Statement

In 2012, several peer-reviewed independent studies were published showing that neonicotinoids had previously undetected routes of exposure affecting bees including through dust, pollen, and nectar and that subnanogram toxicity resulted in failure to return to the hive without immediate lethality, one primary symptom of CCD. Research also showed environmental persistence in agricultural irrigation channels and soil. These reports prompted a formal peer review by the European Food Safety Authority, which stated in January 2013 that some neonicotinoids pose an unacceptably high risk to bees, and identified several data gaps not previously considered. Their review concluded, "A high acute risk to honey bees was identified from exposure via dust drift for the seed treatment uses in maize, oilseed rape and cereals. A high acute risk was also identified from exposure via residues in nectar and/or pollen." Dave Goulson, an author of one of the studies which prompted the EFSA review, has suggested that industry science pertaining to neonicotinoids may have been deliberately deceptive, and the UK Parliament has asked manufacturer Bayer Cropscience to explain discrepancies in evidence they have submitted to an investigation.

Neonicotinoids Banned by European Union

Early in 2013, the European Food Safety Authority issued a declaration that three specific neonicotinoid pesticides pose an acute risk to honey bees, and the European Commission (EC) proposed a two-year ban on them. David Goulson, who led one of the key 2012 studies at the University of Stirling, said the decision "begs the question of what was going on when these chemicals were first approved." The chemical manufacturer Bayer said it was "ready to work with" the EC and member states. In April 2013, the European Union voted for a two-year restriction on neonicotinoid insecticides. The ban will restrict the use of imidacloprid, clothianidin, and thiamethoxam for use on crops that are attractive to bees. Eight nations voted against the motion, including the British government, which argued that the science was incomplete. The ban can be seen as an application of the "precautionary principle", established at the 1992 Rio Conference on the Environment and Development, which advocates that "lack of full scientific certainty shall not be used as a reason for postponing cost-effective measures to prevent environmental degradation."

Initiatives to Ban Neonicotinoids in The United States

In March 2013, professional beekeepers and environmentalists jointly filed a lawsuit against the United States Environmental Protection Agency (EPA) for continuing to allow the use of neonicotinoids in the United States. The suit specifically asks for suspension of clothianidin and thiamethoxam. The lawsuit follows a dramatic die off of bees in the United States, with some beekeepers losing 50% of their hives. The EPA responded to the suit by issuing a report blaming the *Varroa* mite for the decline in bees and claiming the role of neonicotinoids in bee extinction has been overstated.

Pollination

Also in 2013, the Save America's Pollinators Act of 2013 (H.R. 2692) was introduced in Congress. The proposed act asks that neonicotinoids be suspended until a full review of their impacts has occurred. The bill was reintroduced on 4 March 2015 as the Saving America's Pollinators Act (H.R. 1284), where it is currently being debated by the House Subcommittee on Biotechnology, Horticulture, and Research.

Pathogens and Immunodeficiency Theories

Early researchers commented that the pathway of propagation functions in the manner of a contagious disease; however, some sentiment existed that the disorder may involve an immunosuppressive mechanism, potentially linked to "stress" leading to a weakened immune system. Specifically, according to research done in 2007 at the Pennsylvania State University: "The magnitude of detected infectious agents in the adult bees suggests some type of immunosuppression". These researchers initially suggested a connection between *Varroa destructor* mite infestation and CCD, suggesting that a combination of these bee mites, deformed wing virus (which the mites transmit) and bacteria work together to suppress immunity and may be one cause of CCD. Parasites, such as varroa mites *(Varroa destructor)*, honey bee tracheal mites *(Acarapis woodi)*, fungal, bacterial and viral diseases, and kleptoparasites such as small hive beetles *(Aethina tumida)*, are all problems that have been introduced within the last 20 years in the continental U.S., and are faced by beekeepers.

When a colony is dying, for whatever cause, and other healthy colonies are nearby (as is typical in a bee yard), those healthy colonies often enter the dying colony and rob its provisions for their own use. If the dying colony's provisions were contaminated (by natural or man-made toxins), the resulting pattern (of healthy colonies becoming sick when in proximity to a dying colony) might suggest to an observer that a contagious disease is involved. However, it is typical in CCD cases that provisions of dying colonies are not being robbed, suggesting that at least this particular mechanism (toxins being spread via robbing, thereby mimicking a disease) is not involved in CCD. Additional evidence that CCD is an infectious disease came from the following observations: the hives of colonies that had died from CCD could be reused with a healthy colony only if they were first treated with DNA-destroying radiation, and the CCD Working Group report in 2010 indicated that CCD-exhibiting hives tended to occur in proximity to one another within apiaries.

Coumaphos, an organophosphate, is lipophilic, and so accumulates in wax. Increased levels of compound in wax have been shown to decrease survivorship of developing queens.

Varroa Mites

According to a 2007 article, the mite *Varroa destructor* remains the world's most destructive honey bee killer, due in part to the viruses it carries, including deformed wing virus and acute bee paralysis virus, which have both been implicated in CCD. Affliction with *Varroa* mites also tends to weaken the immune system of the bees. Dr. Enesto Guzman, an entomological researcher at the University of Guelph in Canada, studied 413 Ontario bee colonies in 2007–08. About 27% of hives did not survive the winter, and the *Varroa* mite was identified as the cause in 85% of the cases. *Varroa* mites also affect the queen's ability to reproduce which is detrimental to the survival of the hive. As such, *Varroa* mites have been considered as a possible cause of CCD, though not all dying colonies contain these mites.

Varroa destructor on a honey bee host

Israeli Acute Paralysis Virus

In 2004, Israeli acute paralysis virus (IAPV), was discovered in Israel and at one time it was considered the cause of CCD. It was named after the place it was first identified; its place of origin is unknown. In September 2007, results of a large-scale statistical RNA sequencing study of afflicted and unafflicted colonies were reported. RNA from all organisms in a colony was sequenced and compared with sequence databases to detect the presence of pathogens. All colonies were found to be infected with numerous pathogens, but only the IAPV virus showed a significant association with CCD: the virus was found in 25 of the 30 tested CCD colonies, and only in one of the 21 tested non-CCD colonies.

Research in 2009 has found that an indicator for an impaired protein production is common among all bees affected by CCD, a pattern consistent with IAPV infection. It is conjectured that Dicistroviridae, like the IAPV, cause degradation of the ribosomes, which are responsible for protein production of cells, and that this reduced ribosomal function weakens the bees, making them more vulnerable to factors that might not otherwise be lethal.

Nosema

Some have suggested the syndrome may be an inability by beekeepers to correctly identify known diseases such as European foulbrood or the microsporidian fungus *Nosema apis*. The testing and diagnosis of samples from affected colonies (already performed) makes this highly unlikely, as the symptoms are fairly well known and differ from what is classified as CCD. A high rate of *Nosema* infection was reported in samples of bees from Pennsylvania, but this pattern was not reported from samples elsewhere.

When healthy bees are fed pollen filled with fungicides, insecticides and other agriculture chemicals, they are more likely to be infected by *Nosema ceranae*, a parasitic microsporidian fungus associated with widespread death of honey bees. Hives of western honey bees infected with *Nosema ceranae* are wiped out within eight days indicating that CCD may be caused by *N. ceranae*. A research team claim to have ruled out many other potential causes, however, a 2009 survey of US CCD-affected bee populations found only about half of the colonies sampled, both in CCD and control populations, were infected with *N. ceranae*.

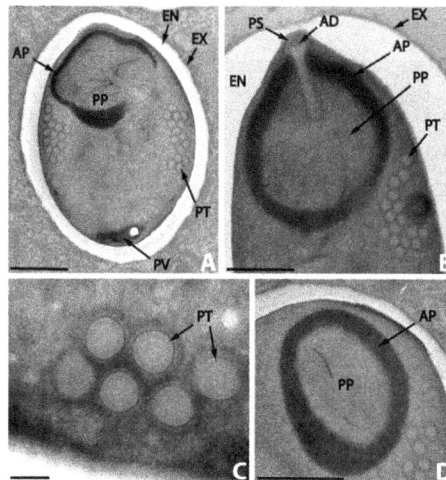

Parasite140019-fig4 Nosema podocotyloidis - Hyperparasitic Microsporidia

The primary antifungal agent used against *Nosema* is fumagillin, which has been used in a German research project to reduce the microsporidian's impact, and is mentioned as a possible remedy by the CCDWG. Higes also claims to have successfully cured colonies with fumagillin. A review of these results described these results as promising, but cautioned "*N. ceranae* may not be to blame for all cases of colony collapse". Various areas in Europe have reported this fungus, but no direct link to CCD has yet been established.

In 2007, *N. ceranae* was reported in a few hives in California. The researcher did not, however, believe this was conclusive evidence of a link to CCD; "We don't want to give anybody the impression that this thing has been solved". A USDA bee scientist has similarly stated, "while the parasite *Nosema ceranae* may be a factor, it cannot be the sole cause. The fungus has been seen before, sometimes in colonies that were healthy".

N. ceranae has been detected in honey bees from several states using PCR of the 16S gene. In New York, *N. ceranae* was detected in 49 counties, and of the 1,200 honey bee samples collected, 528 (44%) were positive for *Nosema*, from which, PCR analysis of 371 spore positive samples revealed 96% were *N. ceranae*, 3% had both *N. ceranae* and *N. apis*, and 1% had *N. apis* only.

Viral and Fungal Combination

A University of Montana and Montana State University team of scientists headed by Jerry Bromenshenk and working with the US Army's Edgewood Chemical Biological Center published a paper in

October 2010 saying that a new DNA virus, invertebrate iridescent virus or IIV6, and the fungus *Nosema ceranae* were found in every killed colony the group studied. In their study, they found neither agent alone seemed deadly, but a combination of the virus and *N. ceranae* was always 100% fatal. Information about the study was released to the public in a front page article in *The New York Times*. A few days later, an article was published in *Fortune Magazine* with the title, "What a scientist didn't tell the New York Times about his study on bee deaths". Professor of entomology at Penn State University James Frazier, who is currently researching the sublethal impact of pesticides on bees, said that while Bromenshenk's study generated some useful data, Bromenshenk has a conflict of interest as CEO of a company developing scanners to diagnose bee diseases. A few months later, the methods used to interpret the mass spectrometry data in the Bromenshenk study were called into question, raising doubts as to whether IIV6 was ever correctly identified in any of the samples examined.

Fungicides

In 2013, researchers collected pollen from hives and fed it to healthy bees. The pollen had an average of nine different pesticides and fungicides. Further, the researchers discovered that bees that ate pollen with fungicides were three times more likely to be infected by parasites. Their study shows that fungicides, thought harmless to bees, may actually play a significant role in CCD. Their research also showed that spraying practices may need to be reviewed because the bees sampled by the authors foraged not from crops, but almost exclusively from weeds and wildflowers, suggesting that bees are more widely exposed to pesticides than thought.

Synhalonia on Phlomis 8

Dennis vanEngelsdorp, an entomologist at the University of Maryland, has been quoted as saying "Fungicides, which we didn't expect to harm insects, seem to have a sub-lethal effect on bee health". He went on further to state this is important because fungicides are not heavily regulated.

Antibiotics and Miticides

Most beekeepers affected by CCD report that they use antibiotics and miticides in their colonies, though the lack of uniformity as to which particular chemicals are used makes it seem unlikely that any single such chemical is involved. However, it is possible that not all such chemicals in use have been tested for possible effects on honey bees, and could therefore potentially be contributing to the CCD phenomenon.

Fluvalinate/Coumaphos

In 2008 high levels of the pesticides fluvalinate and coumaphos were found in samples of wax from hives, as well as lower levels of 70 other pesticides. These chemicals have been used to try to eradicate varroa mites, a bee pest that itself has been thought to be a cause of CCD. A 2009 study confirmed high levels of pesticide residue in hive wax and found an association between the pesticide and reduced bee longevity. *Nosema ceranae*, was found in high concentrations in the majority of the bees tested, even after administering large doses of the antibiotic fumagillin. Maryann Frazier commented, "Pesticides alone have not shown they are the cause of CCD. We believe that it is a combination of a variety of factors, possibly including mites, viruses and pesticides."

Climate Change

Studies in Europe and North America have shown dramatic declines in bee colonies. The U.S. has lost 59% of its bee colonies since 1947, and Europe has lost over 25% since 1985. Determining the causes of bee colony collapse is crucial to the global economy, as bee pollination assists with 9.5% of the global agricultural production worth billions of dollars.

Through ecological modeling and retrospective studies, research has shown a link between bee colony collapse and climate change. Although the shifting weather conditions themselves negatively affect bees, the link between colony collapse and climate change is also closely tied to the interaction between bees' climatic niches and food-plant reductions. Climate change affects the floral environment by stunting flower development and nectar production, which in turn directly impacts colonies' abilities to collect pollen and sustain themselves.

As weather conditions shift due to climate change, bees change their behaviors. When it rains, bees do not go out and during extremely hot weather they try to gather water to keep the colony cool. Climatologists have predicted that the occurrence of extreme weather events (such as intense rainy seasons and prolonged drought) will increase as the climate continues to change. Additionally, in regions that experience increasingly more rain, pollen will be washed away more easily making it more difficult for bees to provide for their colonies. Meanwhile, in environments experiencing prolonged drought, flower environments may dwindle with dry weather. These patterns lead to less suitable and viable environments in which bees can thrive. A British journal published projections of plant diversity loss against spatial sensitivity; the researchers found major species loss in the southern part of the UK forcing bee colonies further north.

Since it can take decades of studying climate change and tracking bee colonies to find correlations between them, many researchers have turned to ecological modeling. In a Brazilian study, researchers investigated the impact that climate change will have on ten endemic bee species in various future climatic scenarios. It is known that moderate temperatures and high relative humidity impacts flight activity and foraging behavior in bee species such as Melipona and Centris. As a result, the scientists predict that in scenarios where temperatures continue to increase and humidity continues to decrease the only suitable climates for bees in Brazil will be in the more mountainous environments. The most optimistic scenario estimated climate change would still lead to a five percent reduction in these species' populations. While this may not seem like a large change, bees are very sensitive to the effects of small population size; small populations can lead to reduced genetic variability and decreased fitness. As a result of this sensitivity, fragmented habitat are less likely to support a viable population and may lead to additional colony collapse.

From these models, researchers also predict that bee populations will start declining most significantly in the southern latitudes with lagging range expansion in the north. Kerr and his colleagues found that as the climate has shifted and bumblebees have been forced to operate outside of their thermal ranges, causing range losses along the species' southern limits. However, the study also found that while other flora and fauna have expanded their northerly ranges to account for the climate shifts, bumblebees have not expanded farther north. The most important findings in this study came from statistical models built to study whether or not these range shifts were due to confounding factors such as pesticide use or changes in land cover. The researchers found that climate change had the most meaningful impact on range shifts and bee distributions. To support these findings, the scientists note that failure for the bumblebee to respond to thermal changes and expand their northern range underscores the idea that bumblebees are susceptible to climate change.

One of the most significant studies on climate change and bees examined the species, B. Disinguendus and B. sylvarum between 2000 and 2006 in the UK. The study showed that as the climate of western Europe warmed, these two bee species experienced range declines and narrower climatic niches pushing them to the fringes of their environments. Geographically and climatically we expect this, as climate change continues bees will move and abandon areas of drought and migrate towards the fringes, as this has already been seen within desert oases. Europe has done significant work in tracking bee colony changes within the regions. Additional longitudinal surveillance needs to be conducted in other regions of the globe among a wider array bee species in order to build the body of research surrounding the impact of climate change on bees.

Bee Rentals and Migratory Beekeeping

Since U.S. beekeeper Nephi Miller first began moving his hives to different areas of the country for the winter of 1908, migratory beekeeping has become widespread in America. Bee rental for pollination is a crucial element of U.S. agriculture, which could not produce anywhere near its current levels with native pollinators alone. U.S. beekeepers collectively earn much more from renting their bees out for pollination than they do from honey production.

Researchers are concerned that trucking colonies around the country to pollinate crops, where they intermingle with other bees from all over, helps spread viruses and mites among colonies.

Moving spring bees from South Carolina to Maine for blueberry pollination

Additionally, such continuous movement and re-settlement is considered by some a strain and disruption for the entire hive, possibly rendering it less resistant to all sorts of systemic disorder.

Selective Commercial Breeding and Lost Genetic Diversity in Industrial Apiculture

Most of the focus on CCD has been toward environmental factors. CCD is a condition recognised for greatest impact in regions of 'industrial' or agricultural use of commercially bred bee colonies. Natural breeding and colony reproduction of wild bees is a complex and highly selective process, leading to a diverse genetic makeup in large populations of bees, both within and between colonies. Genetic diversity through sexual reproduction is a significant evolutionary factor in resistance to parasites and infectious diseases. Many artificially bred species, especially domestic and agricultural species, suffer from lack of genetic variation. resulting in increased risk of hereditable diseases, loss of vitality or vigour, and heightened uniform susceptibility to infectious diseases. There may be an analogy in artificially introduced invasive ants, which displace native species by their ecological release and supercolonies (a manifestation of genetic homogeneity), only to suffer collapse of colonies attributed to lack of genetic diversity. Displaced indigenous species rebounded from residual populations.

Industrial apiculture has adopted simple breeding programs for uniform desired traits, and seasonal transportation of colonies over vast distances causes increased infectious exposures from mixing of these domestic and residual displaced wild populations. Brood incubation conditions may be stressful with respect to deficient nutrition, temperature and other basics. This combination of ecological factors, especially the host factor of loss of genetic variation and hybrid vigor, may account for the apparent multifactorial environmental 'causes' of CCD including concurrent infections.

Malnutrition

In 2007, one of the patterns reported by the CCD Study Group at Pennsylvania State was that all producers in a preliminary survey noted a period of "extraordinary stress" affecting the colonies in question prior to their die-off, most commonly involving poor nutrition and/or drought. This was the only factor that *all* of the cases of CCD had in common in the report; accordingly, there appeared to be at least some significant possibility that the phenomenon was correlated to nutritional stress that may not manifest in healthy, well-nourished colonies. This was similar to the findings of another independent survey done in 2007 in which small-scale beekeeping operations (up to 500 colonies) in several states reported their belief that malnutrition and/or weak colonies was the factor responsible for their bees dying in over 50% of the cases, whether the losses were believed to be due to CCD or not.

Some researchers have attributed the syndrome to the practice of feeding high-fructose corn syrup (HFCS) to supplement winter stores. The variability of HFCS may be relevant to the apparent inconsistencies of results. One European writer has suggested a possible connection with HFCS produced from genetically modified corn.

Other researchers state that colony collapse disorder is mainly a problem of feeding the bees a monoculture diet when they should receive food from a variety of sources/plants. In winter, these bees are given a single food source such as corn syrup (high-fructose or other), sugar and pollen substitute. In summer, they may only pollinate a single crop (e.g., almonds, cherries, or apples). The monoculture diet is attributed to bee rentals and migratory bee keeping. Honey bees are only being introduced to select commercial crops such as corn. These single pollen diets are greatly inferior to mixed pollen diets. However, there are a few pollens that are acceptable for honey bees to be introduced to exclusively, including sweet clover and mustard.

Pupae of honeybee drones in opened cells at both sides of a honeycomb. The drones at the right side are some days older and more developed.

A study published in 2010 found that bees that were fed pollen from a variety of different plant species showed signs of having a healthier immune system than those eating pollen from a single species. Bees fed pollen from five species had higher levels of glucose oxidase than bees fed pollen from one species, even if the pollen had a higher protein content. The authors hypothesised that CCD may be linked to a loss of plant diversity. Researches found a proper diet that does lead to a healthy honey bee population. "The authors recommended a diet containing 1000 ppm potassium, 500 ppm calcium,300 ppm magnesium and 50 ppm each of sodium, zinc, manganese, iron and copper."

The lack of variance in plant pollen does not appear to affect a healthy colony of honey bees however. Once a fungus or parasite invades a colony, research has proven that honey bees introduced to a wider variety of pollen are much more likely to survive longer due to receiving key nutrients and alkaloids like zinc.

A 2013 study found that p-Coumaric acid, which is normally present in honey, assists bees in detoxifying certain pesticides. Its absence in artificial nutrients fed to bees may therefore contribute to CCD.

Electromagnetic Radiation

A study on the non-thermal effects of radio frequency (RF) on honey bees (*Apis mellifera carnica*) reported there were no changes in behavior due to RF exposure from DECT cordless phone base stations operating at 1,880–1,900 MHz however, a later study established that close-range electromagnetic field (EMF) may reduce the ability of bees to return to their hive. In the course of their study, one half of their colonies broke down, including some of their controls which did not have DECT base stations embedded in them. In April 2007, news of this study appeared in various media outlets, beginning with an article in *The Independent*, which stated that the subject of the study included mobile phones and had related them to CCD. Though cellular phones were implicated by other media reports at the time, they were not covered in the study. Researchers involved have since stated that their research did not include findings on cell phones, or their relationship to CCD, and indicated that the *Independent* article had misinterpreted their results and created "a horror story".

A review of 919 peer-reviewed scientific studies investigating the effects of EMF on wildlife, humans and plants included seven studies involving honey bees; six of these reported negative effects from exposure to EMF radiation, but none specifically demonstrated any link to CCD. The review noted that according to one study, when active mobile phones were kept inside beehives, worker bees stopped coming to the hives after 10 days. The same study also found drastic decrease in the egg production of queen bees in these colonies and stated "electromagnetic radiation exposure provides a better explanation for Colony Collapse Disorder (CCD) than other theories". The review authors concluded: "existing literature shows that the EMRs are interfering with the biological systems in more ways than one" and recommended recognising EMF as a pollutant. However, they also noted that "these studies are not representative of the real life situations or natural levels of EMF exposure. More studies need to be taken up to scientifically establish the link, if any, between the observed abnormalities and disorders in bee hives such as Colony Collapse Disorder (CCD)".

Parasitic Phorid Fly

In 2012, a parasitic fly (*Apocephalus borealis*) larva, known to prey on bumble bees and wasps, was found in a test tube containing a dead honey bee believed to have been affected by CCD, possibly indicating the phorid fly may be one cause of CCD. The mature fly lays eggs in the bee's abdomen, which feed on the bee after hatching. Infected bees behave abnormally, foraging at night and gathering around lights like moths. Eventually the bee leaves the colony to die. The phorid fly larvae then emerge from the neck of the bee.

Genetically Modified Crops

In 2008 a meta-analysis of 25 independent studies assessing effects of Bt Cry proteins on honeybee survival (mortality) showed that Bt proteins used in commercialized GE crops to control lepidopteran and coleopteran pests do not negatively impact the survival of honeybee larvae or adults. Additionally, larvae consume only a small percent of their protein from pollen, and there is also a lack of geographic correlation between GM crop locations and regions where CCD occurs.

Management

As of 1 March 2007, the Mid-Atlantic Apiculture Research and Extension Consortium (MAAREC) offered the following tentative recommendations for beekeepers noticing the symptoms of CCD:

1. Do not combine collapsing colonies with strong colonies.

2. When a collapsed colony is found, store the equipment where you can use preventive measures to ensure that bees will not have access to it.

3. If you feed your bees sugar syrup, use Fumagillin.

4. If you are experiencing colony collapse and a secondary infection, such as European Foulbrood, treat the colonies with oxytetracycline, not tylosin.

Another proposed remedy for farmers of pollinated crops is simply to switch from using beekeepers to the use of native bees, such as bumble bees and mason bees. Native bees can be helped to establish themselves by providing suitable nesting locations and some additional crops the bees could use to feed from (e.g. when the pollination season of the commercial crops on the farm has ended).

Beekeeper keeping bees

A British beekeeper successfully developed a strain of bees that are resistant to varroa mites. Russian honey bees also resist infestations of varroa mites but are still susceptible to other factors associated with colony collapse disorder, and have detrimental traits that limit their relevance in commercial apiculture.

In the United Kingdom, a national bee database was set up in March 2009 to monitor colony collapse as a result of a 15% reduction in the bee population that had taken place over the previous two years. In particular, the register, funded by the Department for Environment, Food and Rural Affairs and administered by the National Bee Unit, will be used to monitor health trends and help establish whether the honey industry is under threat from supposed colony collapse disorder. Britain's 20,000 beekeepers have been invited to participate. In October 2010, David Aston of the British Beekeepers' Association stated, "We still do not believe CCD is a cause of colony losses in the UK, however we are continuing to experience colony losses, many if not most of which can be explained. The approach being taken in UK beekeeping is to raise the profile of integrated bee health management, in other words identifying and trying to eliminate factors which reduce the health status of a colony. This incorporates increasing the skill level of beekeepers through training and education, raising the profile of habitat destruction and its effect of forage (nectar and pollen) availability, and of course research on the incidence and distribution of diseases and conditions in the UK together with more applied research and development on providing solutions."

Economic and Ecological Impact

Honey bees are not native to the Americas, therefore their necessity as pollinators in the U.S. and other regions in the Western Hemisphere is limited to strictly agricultural/ornamental uses, as no native plants require honey bee pollination, except where concentrated in monoculture situations—where the pollination need is so great at bloom time that pollinators must be concentrated beyond the capacity of native bees (with current technology).

Bombus argillaceus male 1

The phenomenon is particularly important for crops such as almond growing in California, where honey bees are the predominant pollinator and the crop value in 2011 was $3.6 billion. In 2000, the total U.S. crop value that was wholly dependent on honey bee pollination was estimated to exceed $15 billion. Because of such high demand in pollinators, the cost of renting honey bees has increased significantly, and California's almond industry rents approximately 1.6 million honey bee colonies during the spring to pollinate their crop. Worldwide, honeybees yield roughly $200 billion in pollination services.

They are responsible for pollination of approximately one third of the United States' crop species, including such species as almonds, peaches, apples, pears, cherries, raspberries, blackberries, cranberries, watermelons, cantaloupes, cucumbers, and strawberries. Many, but not all, of these plants can be (and often are) pollinated by other insects in the U.S., including other kinds of bees (e.g., squash bees on cucurbits), but typically not on a commercial scale. While some farmers of a few kinds of native crops do bring in honey bees to help pollinate, none specifically need them, and when honey bees are absent from a region, there is a presumption that native pollinators may reclaim the niche, typically being better adapted to serve those plants (assuming that the plants normally occur in that specific area).

However, even though on a per-individual basis, many other species are actually more efficient at pollinating, on the 30% of crop types where honey bees are used, most native pollinators cannot be mass-utilized as easily or as effectively as honey bees—in many instances they will not visit the plants at all. Beehives can be moved from crop to crop as needed, and the bees will visit many plants in large numbers, compensating via saturation pollination for what they lack in efficiency. The commercial viability of these crops is therefore strongly tied to the beekeeping industry. In China, hand pollination of apple orchards is labor-intensive, time consuming, and costly.

In regions of the Old World where they are indigenous, honeybees (*Apis mellifera*) are among the most important pollinators, vital to sustain natural habitats there in addition to their value for human societies (to sustain food resources). Where honeybee populations decline, there is also a decline in plant populations. In agriculture, some plants are completely dependent on honeybees to pollinate them to produce fruit, while other plants are only dependent on honeybees to enhance their capacity to produce better and healthier fruits. Honeybees also help plants to reduce time between flowering and fruit set, which reduces risk from harmful factors such as pests, diseases, chemicals, weather, etc. Specialist plants that require honeybees will be at more risk if honeybees decline, whereas generalist plants that use other animals as pollinators (or wind pollinating or self-pollinating) will suffer less because they have other sources of pollination.

With that said, honeybees perform some level of pollination of nearly 75% of all plant species directly used for human food worldwide. Catastrophic loss of honeybees could have significant impact, therefore; it is estimated that seven out of the 60 major agricultural crops in North American economy would be lost, and this is only for one region of the world. Farms that have intensive systems (high density of crops) will be impacted the most compared to non-intensive systems (small local gardens that depend on wild bees) because of dependence on honeybees. These types of farms have a high demand for honeybee pollination services, which in the U.S. alone costs $1.25 billion annually. This cost is offset, however, as honeybees as pollinators generate 22.8 to 57 billion Euros globally.

Media

- *Silence of the Bees* (October 2007) is a part of the *Nature* television series and covers several recent investigative discoveries.

- The 2009 documentary *Vanishing of the Bees* pointed to neonicotinoid pesticides as being the most likely culprit, though the experts interviewed concede that no firm data yet exists.

- The 2010 feature-length documentary *Queen of the Sun: What are the bees telling us?* features interviews with beekeepers, scientists, farmers, and philosophers.

- The 2012 documentary, *Nicotine Bees*, argues that neonicotinoid pesticides are principally responsible for Colony Collapse Disorder.

- *More than Honey*, a 2012 documentary, examines the relationship between humans and bees and explores the possible causes of CCD.

Pesticide Formulation

The biological activity of a pesticide, be it chemical or biological in nature, is determined by its active ingredient (AI - also called the *active substance*). Pesticide products very rarely consist of pure technical material. The AI is usually formulated with other materials and this is the product as sold, but it may be further diluted in use. Formulation improves the properties of a chemical for handling, storage, application and may substantially influence effectiveness and safety.

Formulation terminology follows a 2-letter convention: (*e.g.* GR: granules) listed by CropLife International (formerly GIFAP then GCPF) in the *Catalogue of Pesticide Formulation Types* (Monograph 2); Some manufacturers do not follow these industry standards, which can cause confusion for users.

Water-miscible Formulations

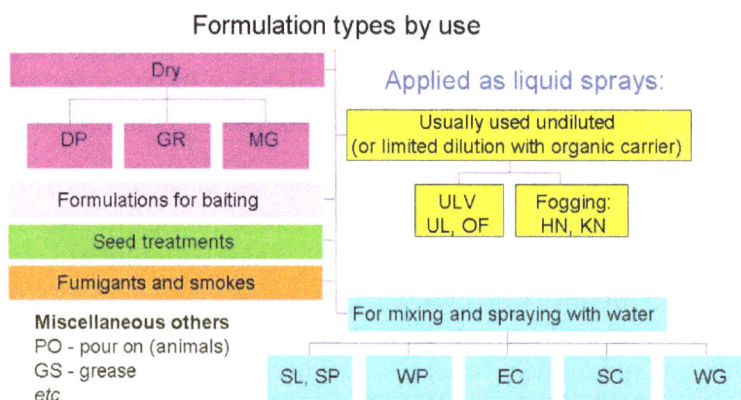

Formulation types by use

Dry

DP   GR   MG

Applied as liquid sprays:

Usually used undiluted
(or limited dilution with organic carrier)

Formulations for baiting

ULV
UL, OF
Fogging:
HN, KN

Seed treatments

Fumigants and smokes

Miscellaneous others
PO - pour on (animals)
GS - grease
etc.

For mixing and spraying with water

SL, SP   WP   EC   SC   WG

By far the most frequently used products are formulations for mixing with water then applying as sprays. Water miscible, older formulations include:

- EC Emulsifiable concentrate

- WP Wettable powder

- SL Soluble (liquid) concentrate

- SP Soluble powder

Newer, non-powdery formulations with reduced or no use of hazardous solvents and improved stability include:

- SC Suspension concentrate

- CS Capsule suspensions

- WG Water dispersible granules

Other Formulations

Other common formulations include granules (GR) and dusts (DP), although for improved safety the latter have been replaced by microgranules (MG *e.g.* for rice farmers in Japan). Specialist formulations are available for ultra-low volume spraying, fogging, fumigation, *etc.* Very occasionally, some pesticides (*e.g.* malathion) may be sold as technical material (TC - which is mostly AI, but also contains small quantities of, usually non-active, by-products of the manufacturing process).

A particularly efficient form of pesticide dose transfer is seed treatment and specific formulations have been developed for this purpose. A number of pesticide bait formulations are available for rodent pest control, *etc.*

In reality many formulation codes are used: AB, AE, AL, AP, BB, BR, CB, CF, CG, CL, CP, CS, DC, DL, DP, DS, DT, EC, ED, EG, EO, ES, EW, FD, FG, FK, FP, FR, FS, FT, FU, FW, GA, GB, GE, GF, GG, GL, GP, GR, GS, GW, HN, KK, KL, KN, KP, LA, LS, LV, MC, ME, MG, MV, OD, OF, OL, OP, PA, PB, PC, PO, PR, PS, RB, SA, SB, SC, SD, SE, SG, SL, SO, SP, SS, ST, SU, TB, TC, TK, TP, UL, VP, WG, WP, WS, WT, XX, ZC, ZE and ZW.

Imidacloprid Effects on Bees

Imidacloprid is a nicotine-derived systemic insecticide, belonging to a group of pesticides called neonicotinoids. Although it is off patent, the primary producer of imidacloprid is the German chemical firm Bayer CropScience. The trade names for imidacloprid include Gaucho, Admire, Merit, Advantage, Confidor, Provado, and Winner. Imidacloprid is a neurotoxin that is selectively toxic to insects relative to vertebrates and most non-insect invertebrates. It acts as an agonist on the postsynaptic nicotinic acetylcholine receptors of motor neurones in insects. This interaction results in convulsions, paralysis, and eventually death of the poisoned insect. It is effective on contact and via stomach action. Because imidacloprid binds much more strongly to insect neuron receptors than to mammal neuron receptors, this insecticide is selectively more toxic to insects than mammals. As a systemic pesticide, imidacloprid translocates or moves readily in the xylem of plants from the soil into the leaves, fruit, flowers, pollen, nectar, and guttation fluid of plants. Bees may be exposed to imidacloprid when they feed on the nectar, pollen, and guttation fluid of imidacloprid-treated plants.

Experts believe that imidacloprid is one of many possible causes of bee decline and the recent bee malady termed colony collapse disorder (CCD). In 2011, according to the United States Department of Agriculture, no single factor alone is responsible for the malady, however honey bees are thought to possibly be affected by neonicotinoid chemicals existing as residues in the nectar and pollen which bees forage on. The scientists studying CCD have tested samples of pollen and have indicated findings of a broad range of substances, including insecticides, fungicides, and herbicides. They note that while the doses taken up by bees are not lethal, they are concerned about possible chronic problems caused by long-term exposure.

In January 2013, the European Food Safety Authority stated that neonicotinoids pose an unacceptably high risk to bees, and that the industry-sponsored science upon which regulatory agencies' claims of safety have relied may be flawed, concluding that, "A high acute risk to honey bees was identified from exposure via dust drift for the seed treatment uses in maize, oilseed rape and cereals. A high acute risk was also identified from exposure via residues in nectar and/or pollen." An author of a *Science* study prompting the EESA review suggested that industry science pertaining to neonicotinoids may have been deliberately deceptive, and the UK Parliament has asked manufacturer Bayer Cropscience to explain discrepancies in evidence they have submitted to an investigation.

April 2013 the EU decided to restrict thiamethoxam and clothianidin along with imidacloprid.

History

Regulatory and Usage History

Imidacloprid was first registered in the United Kingdom in 1993 and in the United States and France in 1994. In the mid to late 1990s, French beekeepers reported a significant loss of bees, which they attributed to the use of imidacloprid. In 1999, the French Minister of Agriculture suspended the use of imidacloprid on sunflower seeds and appointed a team of expert scientists to examine the impact of imidacloprid on bees. In 2003, this panel, referred to as the Comité Scientifique et Technique (CST, or Scientific and Technical Committee) issued a 108-page report, which concluded that imidacloprid poses a significant risk to bees. In 2004, the French Minister of Agriculture suspended the use of imidacloprid as a seed treatment for maize (corn). Despite these bans, colony collapse disorder still is occurring.

Like most insecticides, imidacloprid is highly toxic to bees, with a contact acute LD50 = 0.078µg a.i./bee and an acute oral LD50 = 0.0039µg a.i./bee. Imidacloprid was first widely used in the United States in 1996 as it replaced 3 broad classes of insecticides. In 2006, U.S. commercial migratory beekeepers reported sharp declines in their honey bee colonies. This has happened in the past; however, unlike previous losses, adult bees were abandoning their hives. Scientists named this phenomenon colony collapse disorder (CCD). Reports show that beekeepers in most states have been affected by CCD. Although no single factor has been identified as causing CCD, the United States Department of Agriculture (USDA), in their progress report on CCD, stated that CCD may be "a syndrome caused by many different factors, working in combination or synergistically."

In a British parliamentary inquiry in 2012, the Environmental Audit Committee accused European regulators of ignoring evidence of imidacloprid risk to bees. The committee said that imidacloprid data available in the regulators' own assessment report shows "unequivocally that imidacloprid breaks down very slowly in soil, so that concentrations increase significantly year after year with repeated use, accumulating to concentrations very likely to cause mass mortality in most soil-dwelling animal life." The committee submitted a lengthy list of failings in current regulations including concerns that current regulations were set up for pesticide sprays, not systemic chemicals like imidacloprid that is used to treat seeds. They also expressed concern that only effects on honeybees have been considered despite the fact that 90% of pollination is carried out by different species, such as bumblebees, butterflies, moths and other insects. The environment minister responded saying that he is presently "...satisfied that the [European regulatory system] is working properly."

Research History

Dave Goulson from the University of Stirling has found that trivial effects due to imidacloprid in lab and greenhouse experiments can translate into large effects in the field. The research found that bees consuming the pesticide suffered an 85% loss in the number of queens their nests produced, and a doubling of the number of bees who failed to return from food foraging trips.

Researchers from the Harvard School of Public Health write that new research provides "convincing evidence" of the link between imidacloprid and the phenomenon known as Colony Collapse Disorder. Lead author of the study, Chensheng (Alex) Lu, stated that experiments showed a dose of

20 parts per billion of imidacloprid (less than the concentrations bees would encounter while foraging in sprayed crops), was enough to lead to Colony Collapse Disorder in 94% of colonies within 23 weeks. The hives were nearly empty and the researchers did not find signs of the *Nosema* virus or *Varroa* mites. The researchers proposed two possible sources of bees' exposure to imidacloprid. The first is through the nectar of plants sprayed with the pesticide itself, which is predicted by researchers at the University of Stirling, U.K., to have widespread impacts as imidacloprid is registered for use on over 140 crops in at least 120 countries. The second is through the high-fructose corn-syrup that most bee-keepers in the United States use to feed their bees. Since application of imidacloprid to corn in the United States began in 2005 cases of Colony Collapse Disorder have grown significantly: from losses of 17% to 20% throughout the 1990s to somewhere between 30% and 90% of colonies in the United States since 2006.

In May 2012, researchers at the University of San Diego released a study that showed that honey bees treated with a small dose of imidacloprid, comparable to what they would receive in nectar and formerly considered a safe amount, became "picky eaters," refusing nectars of lower sweetness and preferring to feed only on sweeter nectar. It was also found that bees exposed to imidacloprid performed the "waggle dance," the movements that bees use to inform hive mates of the location of foraging plants, at a lesser rate.

Also in 2012, USDA researcher Jeff Pettis published the results of his study, which showed that bees treated with sub-lethal or low levels of imidacloprid had higher rates of infection with the pathogen *Nosema* than untreated bees. His research confirmed that done by Alaux (2010) and Vidau (2011), who found that interactions between *Nosema* and neonicotinoids weakened bees and led to increased mortality.

Toxicity of Imidacloprid to Bees

Acute

Imidacloprid is one of the most toxic insecticides to the western honeybee, *Apis mellifera*. The toxicity of imidacloprid to *Apis mellifera* differs from most insecticides in that it is more toxic orally than by contact. The contact acute LD_{50} is 0.024 µg a.i./bee (micrograms of active ingredient per bee). The acute oral LD_{50} ranges from 0.005 µg a.i./bee to 0.07 µg a.i./bee, which makes imidacloprid more toxic to the bees than the organophosphate dimethoate (oral LD_{50} 0.152 µg/bee) or the pyrethroid cypermethrin (oral LD_{50} 0.160 µg/bee). Other insecticides that are equally or more toxic than imidacloprid include spinosad, emamectin benzoate, fipronil, and the neonicotinoids clothianidin, thiamethoxam, and dinotefuran.

Sublethal

The majority of studies that measure toxicity of pesticides to *Apis mellifera* honeybees focus on estimating the lethal dose (LD50) in acute toxicity tests to adult honeybees. This is only a partial measure of the harmful effects that pesticides can have on bees. For a complete analysis of the impact of pesticides to bees, sublethal effects should be considered.

Dozens of research articles have been published in peer-reviewed journals, which show sublethal effects to adult bees exposed to low levels of imidacloprid. In these studies, sub-lethal doses of

1-24µg/kg and 0.1 - 20 ng/bee have been shown to impair navigation, foraging behavior, feeding behavior, and olfactory learning performance in honeybees (*Apis mellifera*). Other studies examining higher levels of imidacloprid (50 - 500 ppb) also found that imidacloprid decreases foraging activity and affects bee mobility and communication capacity.

A 2012 in situ study sought to recreate hypothesized conditions of the initial outbreak of CCD in 2006/2007 by feeding honey bees high fructose corn syrup (HFCS) that the researchers laced with varying sub-lethal amounts of imidacloprid assumed to have been present in HFCS feed at the time. All but one of the colonies exposed to imidacloprid perished between 13 and 23 weeks post imidacloprid dosing, providing evidence that long-term sub-lethal exposure to the neonicotinoid causes honey bees to exhibit symptoms consistent with CCD months after exposure.

Chronic

In 10-day chronic feeding studies with honeybees (*Apis mellifera*), 50% mortality was reached at levels between 0.1 and 10 ug/kg imidacloprid. Other chronic toxicity studies conducted by Moncharmont et al. (2003) and Decourtye et al. (1999) have demonstrated chronic NOAEC values of <4 ppb and 4 ppb, respectively in honeybees. In bumble bees, Mommaerts et al. (2009) demonstrated a LOAEC of 10 ppb for imidacloprid.

Many tunnel and field studies have been conducted to show the potential effects of imidacloprid in the natural environment however most of these field studies have design and implementation deficiencies, which make them difficult to interpret and use.

Synergistic Effects

In 2012, researchers announced findings that sublethal exposure to imidacloprid rendered honey bees significantly more susceptible to infection by the fungus *Nosema*, thereby suggesting a potential link to CCD. Two research teams led by Jeff Pettis at the U.S. Department of Agriculture and Cedric Alaux at INRA/France have demonstrated that interactions between the pathogen *Nosema* and imidacloprid significantly weaken the immune systems of honeybees (*Apis mellifera*). In their research, Alaux et al. (2010) found that bees infected with *Nosema* and exposed to 0.7 ug/kg imidacloprid had an increased rate of mortality compared to the controls. The combination of *Nosema* and imidacloprid also significantly decreased the activity of glucose oxidase, an important enzyme that allows the bees to sterilize their colony and brood food. Without this enzyme, bees can become more susceptible to infections by pathogens. Both the USDA study and the INRA study demonstrate that a combination of stressors (pesticides and pathogens) may be responsible for the recent high level of bee losses.

Other Studies

Bayer CropScience studies show that the maximum dose of imidacloprid for which no adverse effects were observed in bees is 20 ppb. Since Bayer claims that residue levels are usually below 5 ppb in pollen and nectar, they contend that imidacloprid poses a negligible risk to bees.

Gerard Eyries, marketing manager for Bayer's agricultural division in France, states that studies confirm that imidacloprid leaves a small residue in nectar and pollen, but there is no evidence of a link with the drop in France's bee population, adding, "It is impossible to have zero residue. What is important is to know whether the very tiny quantities which have been found have a negative effect on bees." He also added that the product was sold in 70 countries with no reported side effects.

Other independent studies have indicated that imidacloprid residues in plants can be higher:

- 10 to 20 ppb in upper leaves
- 100 to 200 ppb in other leaves
- 1.5 ppb in nectar
- 2 to 70 ppb in pollen

Uncertainties

It is important to note that the majority of studies conducted on pollinators have been performed in adult honeybees (*Apis mellifera*). Very few studies have been conducted on wild bees, most of which are solitary and raise their young in burrows and small colonies. There are also few studies that have been conducted on brood, larvae, or the queen, making it difficult to determine the impact of pesticides on different members of the colony and life stages of the bee. Although a number of field and semi-field studies have been conducted on imidacloprid and bees, these studies have design and implementation deficiencies, which make them unusable. Thus, the chronic effects of imidacloprid in the field are still unknown.

Media Portrayal

In October 2009, a documentary film, *Vanishing of the Bees*, was released in theatres in the UK. The film interviewed a number of experts in connection with CCD and suggested a link does exist between neonicotinoid pesticides and CCD. However, the experts interviewed conceded no firm scientific data yet exist. Industry-sponsored studies appear to be inconsistent with those produced by independent scientists. Until 2009 regulatory agencies still did not have conclusive data to determine the effects of imidacloprid on bee colonies.

In February 2010, the documentary film *Nicotine Bees* was released. This film analyzes the possible factors contributing to the large bee die-offs worldwide and concludes that the large use of neonicotinoids is the most probable cause of the recent bee die-offs.

References

- Dennis vanEngelsdorp; Diana Cox-Foster; Maryann Frazier; Nancy Ostiguy; Jerry Hayes (5 January 2006). "Colony Collapse Disorder Preliminary Report" (PDF). Mid-Atlantic Apiculture Research and Extension Consortium (MAAREC) – CCD Working Group. p. 22. Retrieved 2007-04-24.
- Wines, Michael (28 March 2013). "Mystery Malady Kills More Bees, Heightening Worry on Farms". New York Times. Retrieved 31 March 2013.
- Robyn M. Underwood; Dennis van Engelsdorp. "Colony Collapse Disorder: Have We Seen This Before?". The Pennsylvania State University, Department of Entomology. Retrieved 2010-05-02.
- USDA (2014). "Yearly survey shows better results for pollinators, but losses remain significant Release No. 0088.14". USDA. Retrieved 21 August 2015.

- Alison Benjamin (2 May 2010). "Fears for crops as shock figures from America show scale of bee catastrophe | Environment | The Observer". London: Guardian. Retrieved 2010-06-22.

- Le Conte, Yves; Ellis, Marion; Ritter, Wolfgang (2010). "Varroa mites and honey bee health: can Varroa explain part of the colony losses?" (PDF). Apidologie. 41 (3): 353–363. doi:10.1051/apido/2010017. Retrieved 28 May 2014.

- Becher, Matthias A.; Osborne, Juliet L.; Thorbek, Pernille; Kennedy, Peter J.; Grimm, Volker (2013). "Towards a systems approach for understanding honeybee decline: a stocktaking and synthesis of existing models". Journal of Applied Ecology. 50 (4): 868–880. doi:10.1111/1365-2664.12112. PMID 24223431. Retrieved 28 May 2014.

- USDA (17 October 2012). Report on the National Stakeholders Conference on Honey Bee Health National Honey Bee Health Stakeholder Conference Steering Committee (PDF) (Report). Retrieved 4 June 2014.

- Genersch, Elke (2010). "Honey bee pathology: current threats to honey bees and beekeeping" (PDF). Appl Microbiol Biotechnol. 87 (1): 87–97. doi:10.1007/s00253-010-2573-8. PMID 20401479. Retrieved 28 May 2014.

- Smith, Kristine M.; Loh, Elizabeth H.; Rostal, Melinda K.; Zambrana-Torrelio, Carlos M.; Mendiola, Luciana; Daszak, Peter (2013). "Pathogens, Pests, and Economics: Drivers of Honey Bee Colony Declines and Losses". EcoHealth. 4 (4): 434–445. doi:10.1007/s10393-013-0870-2. Retrieved 28 May 2014.

- "CCD Steering Committee, Colony Collapse Disorder Progress Report (US Department of Agriculture, Washington, DC, 2009)" (PDF). Retrieved 2010-06-22.

9

Concepts of Plant Pathology

Plant pathology studies the pathological and environmental conditions that plague plants, disease cycles, plant resistance to disease etc. Plant disease resistance is exhibited either through in-built chemicals and anatomy or immune responses after exposure to diseases. This chapter examines plant pathology, plant disease resistance and pest control.

Plant Pathology

Plant pathology (also phytopathology) is the scientific study of diseases in plants caused by pathogens (infectious organisms) and environmental conditions (physiological factors). Organisms that cause infectious disease include fungi, oomycetes, bacteria, viruses, viroids, virus-like organisms, phytoplasmas, protozoa, nematodes and parasitic plants. Not included are ectoparasites like insects, mites, vertebrate, or other pests that affect plant health by consumption of plant tissues. Plant pathology also involves the study of pathogen identification, disease etiology, disease cycles, economic impact, plant disease epidemiology, plant disease resistance, how plant diseases affect humans and animals, pathosystem genetics, and management of plant diseases.

Overview

Control of plant diseases is crucial to the reliable production of food, and it provides significant reductions in agricultural use of land, water, fuel and other inputs. Plants in both natural and cultivated populations carry inherent disease resistance, but there are numerous examples of devastating plant disease impacts, as well as recurrent severe plant diseases. However, disease control is reasonably successful for most crops. Disease control is achieved by use of plants that have been bred for good resistance to many diseases, and by plant cultivation approaches such as crop rotation, use of pathogen-free seed, appropriate planting date and plant density, control of field moisture, and pesticide use. Across large regions and many crop species, it is estimated that diseases typically reduce plant yields by 10% every year in more developed settings, but yield loss to diseases often exceeds 20% in less developed settings. Continuing advances in the science of plant pathology are needed to improve disease control, and to keep up with changes in disease pressure caused by the ongoing evolution and movement of plant pathogens and by changes in agricultural practices. Plant diseases cause major economic losses for farmers worldwide. The Food and Agriculture Organization estimates indeed that pests and diseases are responsible for about 25% of crop loss. To solve this issue, new methods are needed to detect diseases and pests early, such as novel sensors that detect plant odours and spectroscopy and biophotonics that are able to diagnostic plant health and metabolism.

Plant Pathogens

Fungi

Most phytopathogenic fungi belong to the Ascomycetes and the Basidiomycetes.

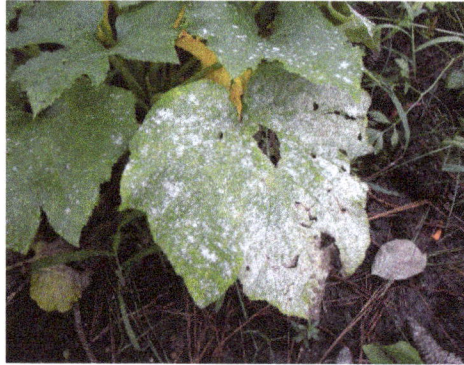

Powdery mildew, a biotrophic fungus

The fungi reproduce both sexually and asexually via the production of spores and other structures. Spores may be spread long distances by air or water, or they may be soilborne. Many soil inhabiting fungi are capable of living saprotrophically, carrying out the part of their life cycle in the soil. These are known as facultative saprotrophs.

Fungal diseases may be controlled through the use of fungicides and other agriculture practices. However, new races of fungi often evolve that are resistant to various fungicides.

Biotrophic fungal pathogens colonize living plant tissue and obtain nutrients from living host cells. Necrotrophic fungal pathogens infect and kill host tissue and extract nutrients from the dead host cells. See the powdery mildew and rice blast images, below.

Rice blast, caused by a necrotrophic fungus

Significant fungal plant pathogens include:

Ascomycetes

- *Fusarium* spp. (causal agents of Fusarium wilt disease)

- *Thielaviopsis* spp. (causal agents of: canker rot, black root rot, *Thielaviopsis* root rot)

- *Verticillium* spp.

- *Magnaporthe grisea* (causal agent of rice blast)

- *Sclerotinia sclerotiorum* (causal agent of cottony rot)

Basidiomycetes

- *Ustilago* spp. (causal agents of smut)

- *Rhizoctonia* spp.

- *Phakospora pachyrhizi* (causal agent of soybean rust)

- *Puccinia* spp. (causal agents of severe rusts of virtually all cereal grains and cultivated grasses)

- *Armillaria* spp. (the so-called honey fungus species, which are virulent pathogens of trees and produce edible mushrooms)

Fungus-like Organisms

Oomycetes

The oomycetes are not true fungi but are fungus-like organisms. They include some of the most destructive plant pathogens including the genus *Phytophthora*, which includes the causal agents of potato late blight and sudden oak death. Particular species of oomycetes are responsible for root rot.

Despite not being closely related to the fungi, the oomycetes have developed very similar infection strategies. Oomycetes are capable of using effector proteins to turn off a plant's defenses in its infection process. Plant pathologists commonly group them with fungal pathogens.

Significant oomycete plant pathogens

- *Pythium* spp.

- *Phytophthora* spp., including the causal agent of the Great Irish Famine (1845–1849)

Phytomyxea

Some slime molds in Phytomyxea cause important diseases, including club root in cabbage and its relatives and powdery scab in potatoes. These are caused by species of *Plasmodiophora* and *Spongospora*, respectively.

Bacteria

Most bacteria that are associated with plants are actually saprotrophic and do no harm to the plant itself. However, a small number, around 100 known species, are able to cause disease. Bacterial diseases are much more prevalent in subtropical and tropical regions of the world.

Crown gall disease caused by Agrobacterium

Most plant pathogenic bacteria are rod-shaped (bacilli). In order to be able to colonize the plant they have specific pathogenicity factors. Five main types of bacterial pathogenicity factors are known: uses of cell wall–degrading enzymes, toxins, effector proteins, phytohormones and exopolysaccharides.

Pathogens such as *Erwinia* species use cell wall–degrading enzymes to cause soft rot. *Agrobacterium* species change the level of auxins to cause tumours with phytohormones. Exopolysaccharides are produced by bacteria and block xylem vessels, often leading to the death of the plant.

Bacteria control the production of pathogenicity factors via quorum sensing.

Vitis vinifera with "Ca. Phytoplasma vitis" infection

Significant bacterial plant pathogens:

- Burkholderia

- Proteobacteria

    o *Xanthomonas* spp.

    o *Pseudomonas* spp.

- Pseudomonas syringae pv. tomato causes tomato plants to produce less fruit, and it "continues to adapt to the tomato by minimizing its recognition by the tomato immune system."

Phytoplasmas ('Mycoplasma-like Organisms') and Spiroplasmas

Phytoplasma and *Spiroplasma* are a genre of bacteria that lack cell walls and are related to the mycoplasmas, which are human pathogens. Together they are referred to as the mollicutes. They also tend to have smaller genomes than most other bacteria. They are normally transmitted by sap-sucking insects, being transferred into the plants phloem where it reproduces.

Tobacco mosaic virus

Viruses, Viroids and Virus-like Organisms

There are many types of plant virus, and some are even asymptomatic. Under normal circumstances, plant viruses cause only a loss of crop yield. Therefore, it is not economically viable to try to control them, the exception being when they infect perennial species, such as fruit trees.

Most plant viruses have small, single-stranded RNA genomes. However some plant viruses also have double stranded RNA or single or double stranded DNA genomes. These genomes may encode only three or four proteins: a replicase, a coat protein, a movement protein, in order to allow cell to cell movement through plasmodesmata, and sometimes a protein that allows transmission by a vector. Plant viruses can have several more proteins and employ many different molecular translation methods.

Plant viruses are generally transmitted from plant to plant by a vector, but mechanical and seed transmission also occur. Vector transmission is often by an insect (for example, aphids), but some fungi, nematodes, and protozoa have been shown to be viral vectors. In many cases, the insect and virus are specific for virus transmission such as the beet leafhopper that transmits the curly top virus causing disease in several crop plants.

Nematodes

Nematodes are small, multicellular wormlike animals. Many live freely in the soil, but there are some species that parasitize plant roots. They are a problem in tropical and subtropical regions of the world, where they may infect crops. Potato cyst nematodes (*Globodera pallida* and *G. rostochiensis*) are widely distributed in Europe and North and South America and cause $300 million worth of damage in Europe every year. Root knot nematodes have quite a large host range, whereas cyst nematodes tend to be able to infect only a few species. Nematodes are able to cause radical changes in root cells in order to facilitate their lifestyle.

Root-knot nematode galls

Protozoa and Algae

There are a few examples of plant diseases caused by protozoa (e.g., *Phytomonas*, a kinetoplastid). They are transmitted as zoospores that are very durable, and may be able to survive in a resting state in the soil for many years. They have also been shown to transmit plant viruses.

When the motile zoospores come into contact with a root hair they produce a plasmodium and invade the roots.

Some colourless parasitic algae (e.g., *Cephaleuros*) also cause plant diseases.

Parasitic Plants

Parasitic plants such as mistletoe and dodder are included in the study of phytopathology. Dodder, for example, is used as a conduit either for the transmission of viruses or virus-like agents from a host plant to a plant that is not typically a host or for an agent that is not graft-transmissible.

Common Pathogenic Infection Methods

- Cell wall-degrading enzymes: These are used to break down the plant cell wall in order to release the nutrients inside.

- Toxins: These can be non-host-specific, which damage all plants, or host-specific, which cause damage only on a host plant.

- Effector proteins: These can be secreted into the extracellular environment or directly into the host cell, often via the Type three secretion system. Some effectors are known to suppress host defense processes. This can include: reducing the plants internal signaling mechanisms or reduction of phytochemicals production. Bacteria, fungus and oomycetes are known for this function.

Physiological Plant Disorders

Significant abiotic disorders can be caused by:

Natural

Drought

Frost damage and breakage by snow and hail

Flooding and poor drainage

Nutrient deficiency

Salt deposition and other soluble mineral excesses (e.g., gypsum)

Wind (windburn and breakage by hurricanes and tornadoes)

Lightning and wildfire (also often man-made)

Man-made (arguably not abiotic, but usually regarded as such)

Soil compaction

Pollution of air, soil, or both

Salt from winter road salt application or irrigation

Herbicide over-application

Poor education and training of people working with plants (e.g. lawnmower damage to trees)

Vandalism

Orchid leaves with viral infections

Disease Resistance

Management

Quarantine

A diseased patch of vegetation or individual plants can be isolated from other, healthy growth. Specimens may be destroyed or relocated into a greenhouse for treatment or study. Another option is to avoid the introduction of harmful nonnative organisms by controlling all human traffic and activity (e.g., AQIS), although legislation and enforcement are crucial in order to ensure lasting effectiveness.

Cultural

Farming in some societies is kept on a small scale, tended by peoples whose culture includes farming traditions going back to ancient times. (An example of such traditions would be lifelong training in techniques of plot terracing, weather anticipation and response, fertilization, grafting, seed care, and dedicated gardening.) Plants that are intently monitored often benefit from not only active external protection but also a greater overall vigor. While primitive in the sense of being the most labor-intensive solution by far, where practical or necessary it is more than adequate.

Plant resistance

Sophisticated agricultural developments now allow growers to choose from among systematically cross-bred species to ensure the greatest hardiness in their crops, as suited for a particular region's pathological profile. Breeding practices have been perfected over centuries, but with the advent of genetic manipulation even finer control of a crop's immunity traits is possible. The engineering of food plants may be less rewarding, however, as higher output is frequently offset by popular suspicion and negative opinion about this "tampering" with nature.

Chemical

Many natural and synthetic compounds can be employed to combat the above threats. This method works by directly eliminating disease-causing organisms or curbing their spread; however, it has been shown to have too broad an effect, typically, to be good for the local ecosystem. From an economic standpoint, all but the simplest natural additives may disqualify a product from "organic" status, potentially reducing the value of the yield.

Biological

Crop rotation may be an effective means to prevent a parasitic population from becoming well-established, as an organism affecting leaves would be starved when the leafy crop is replaced by a tuberous type, etc. Other means to undermine parasites without attacking them directly may exist.

Integrated

The use of two or more of these methods in combination offers a higher chance of effectiveness.

Timeline of Plant Pathology

300–286 BC Theophrastus, father of botany, wrote and studied diseases of trees, cereals and legumes

1665 Robert Hooke illustrates a plant-pathogenic fungal disease, rose rust

1675 Anton van Leeuwenhouek invents the compound microscope, in 1683 describes bacteria seen with the microscope

1729 Pier Antonio Micheli, father of mycology, observes spores for the first time, conducts germination experiments

1755 Tillet reports on treatment of seeds

1802 Lime sulfur first used to control plant disease

1845–1849 Potato late blight epidemic in Ireland

1853 Heinrich Anton de Bary father of modern mycology, establishes that fungi are the cause, not the result, of plant diseases, publishes "Untersuchungen uber die Brandpilze"

1858 Julius Kühn publishes "Die Krankheiten der Kultergewachse"

1865 M. Planchon discovers a new species of *Phylloxera*, which was named *Phylloxera vastatrix.*

1868–1882 Coffee rust epidemic in Sri Lanka

1875 Mikhail Woronin identified the cause of clubroot as a "plasmodiophorous organism" and gave it the name *Plasmodiophora brassicae*

1876 *Fusarium oxysporum* f.sp. *cubense*, responsible for Panama disease, discovered in bananas in Australia

1878–1885 Downy mildew of grape epidemic in France

1879 Robert Koch establishes germ theory: diseases are caused by microorganisms

1882 *Lehrbuch der Baumkrankheiten* (*Textbook of Diseases of Trees*), by Robert Hartig, is published in Berlin, the first textbook of forest pathology.

1885 Bordeaux mixture introduced by Pierre-Marie-Alexis Millardet to control downy mildew on grape

1885 Experimental proof that bacteria can cause plant diseases: "Erwinia amylovora" and fire blight of apple

1886–1898 Recognition of plant viral diseases: Tobacco mosaic virus

1889 Introduction of hot water treatment of seed for disease control by Jensen

1902 First chair of plant pathology established, in Copenhagen

1904 Mendelian inheritance of cereal rust resistance demonstrated

1907 First academic department of plant pathology established, at Cornell University

1908 American Phytopathological Society founded

1910 Panama disease reaches Western Hemisphere

1911 Scientific journal *Phytopathology* founded

1925 Panama disease reaches every banana-growing country in the Western Hemisphere

1951 European and Mediterranean Plant Protection Organization (EPPO) founded

1967 Recognition of plant pathogenic mycoplasma-like organisms

1971 T. O. Diener discovers viroids, organisms smaller than viruses

The historical landmarks in plant pathology are taken from unless otherwise noted.

Plant Disease Resistance

Plant disease resistance protects plants from pathogens in two ways: mechanisms and by infection-induced responses of the immune system. Relative to a susceptible plant, disease resistance is the reduction of pathogen growth on or in the plant, while the term disease tolerance describes plants that exhibit little disease damage despite substantial pathogen levels. Disease outcome is determined by the three-way interaction of the pathogen, the plant and the environmental conditions (an interaction known as the disease triangle).

Defense-activating compounds can move cell-to-cell and systemically through the plant vascular system. However, plants do not have circulating immune cells, so most cell types exhibit a broad suite of antimicrobial defenses. Although obvious *qualitative* differences in disease resistance can be observed when multiple specimens are compared (allowing classification as "resistant" or "susceptible" after infection by the same pathogen strain at similar inoculum levels in similar environments), a gradation of *quantitative* differences in disease resistance is more typically observed between plant strains or genotypes. Plants consistently resist certain pathogens but succumb to others; resistance is usually pathogen species- or pathogen strain-specific.

Background

Plant disease resistance is crucial to the reliable production of food, and it provides significant reductions in agricultural use of land, water, fuel and other inputs. Plants in both natural and cultivated populations carry inherent disease resistance, but this has not always protected them.

The late blight Irish potato famine of the 1840s was caused by the oomycete Phytophthora infestans. The world's first mass-cultivated banana cultivar Gros Michel was lost in the 1920s to Panama disease caused by the fungus Fusarium oxysporum. The current wheat stem, leaf, and yellow stripe rust epidemics spreading from East Africa into the Indian subcontinent are caused by rust fungi Puccinia graminis and P. striiformis. Other epidemics include Chestnut blight, as well as recurrent severe plant diseases such as Rice blast, Soybean cyst nematode, Citrus canker.

Plant pathogens can spread rapidly over great distances, vectored by water, wind, insects, and humans. Across large regions and many crop species, it is estimated that diseases typically reduce plant yields by 10% every year in more developed nations or agricultural systems, but yield loss to diseases often exceeds 20% in less developed settings, an estimated 15% of global crop production.

However, disease control is reasonably successful for most crops. Disease control is achieved by use of plants that have been bred for good resistance to many diseases, and by plant cultivation approaches such as crop rotation, pathogen-free seed, appropriate planting date and plant density, control of field moisture and pesticide use.

Viral Disease Common Mechanisms

Pre-formed Structures and Compounds

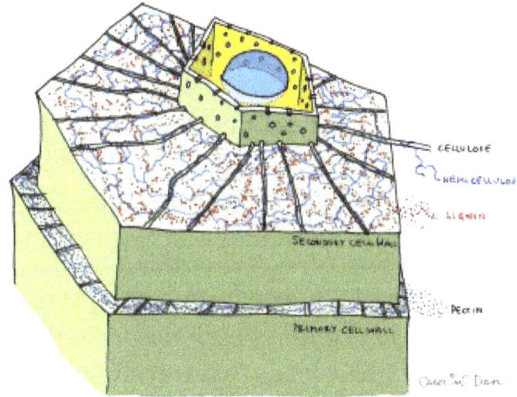

secondary plant wall

- Plant cuticle/surface

- Plant cell walls

- Antimicrobial chemicals (for example: glucosides, saponins)

- Antimicrobial proteins

- Enzyme inhibitors

- Detoxifying enzymes that break down pathogen-derived toxins

- Receptors that perceive pathogen presence and activate inducible plant defences

Inducible Post-infection Plant Defenses

- Cell wall reinforcement (callose, lignin, suberin, cell wall proteins)

- Antimicrobial chemicals, including reactive oxygen species such as hydrogen peroxide or peroxynitrite, or more complex phytoalexins such as genistein or camalexin

- Antimicrobial proteins such as defensins, thionins, or PR-1

- Antimicrobial enzymes such as chitinases, beta-glucanases, or peroxidases

- Hypersensitive response - a rapid host cell death response associated with defence mediated by "Resistance genes."(Bryant, Tracy 2008).

Variable Resistance

Even in susceptible plants to obligate parasites the tissue resistance changes due to ontogeny and to influence of external conditions. This resistance can be measured by the value of redox potential of electron carriers, which is produced in the plant by enzymatic reactions associated with respiration. Electron carriers are water-soluble and are not oxidized by air oxygen. There is not free oxygen in the cells, all the oxidizations and reductions take place enzymatically. These reactions are highly specific for the plant species. The host and parasite have different electron carriers.

Immune System

The plant immune system consists of two interconnected tiers of receptors, one outside and one inside the cell. Both systems sense the intruder, respond to the intrusion and optionally signal to the rest of the plant and sometimes to neighboring plants that the intruder is present. The two systems detect different types of pathogen molecules and classes of plant receptor proteins

The first tier is primarily governed by pattern recognition receptors that are activated by recognition of evolutionarily conserved pathogen or microbial–associated molecular patterns (PAMPs or MAMPs, here P/MAMP). Activation of PRRs leads to intracellular signaling, transcriptional reprogramming, and biosynthesis of a complex output response that limits colonization. The system is known as PAMP-Triggered Immunity (PTI)"(JonesDangl2010).

The second tier (again, primarily), effector-triggered immunity (ETI), consists of another set of receptors, (nucleotide-binding)They operate within the cell, encoded by R genes. The presence of specific pathogen "effectors" activates specific NLR proteins that limit pathogen proliferation.

Receptor responses include ion channel gating, oxidative burst, cellular redox changes, or protein kinase cascades that directly activate cellular changes (such as cell wall reinforcement or antimicrobial production), or activate changes in gene expression that then elevate other defensive responses

Plant immune systems show some mechanistic similarities with the immune systems of insects and mammals, but also exhibit many plant-specific characteristics. Plants can sense the presence of pathogens and the effects of infection via activated by touch]. Rice Universi

PAMP-triggered Immunity

PAMP-Triggered Immunity conserved molecules that inhabit multiple pathogen genera are classified as MAMPs by some researchers. The defenses induced by MAMP perception are sufficient to repel most pathogens. However, pathogen effector proteins are adapted to suppress basal defenses such as PTI

Effector Triggered Immunity

Effector Triggered Immunity (ETI) is activated by the presence of pathogen effectors. The ETI immune response is reliant on R genes, and is activated by specific pathogen strains. As with PTI, many specific examples of apparent ETI violate commoMost plant immune systems carry a repertoire of 100-600 different R genes that mediate resistance to various virus .Plant ETI often cause an apoptotic hypersensitive response.(Odds Rathjen2010).

R Genes and R Proteins

Plants have evolved R genes (resistance genes) whose products allow recognition of specific pathogen effectors, either through direct binding or by recognition of the effector's alteration of a host protein. These virulence factors drove co-evolution of plant resistant genes to combat the pathogens' Avr (avirulent) genes. Many R genes encode NB-LRR proteins (nucleotide-binding/leucine-rich repeat domains, also known as NLR proteins or STAND proteins, among other names).

R gene products control a broad set of disease resistance responses whose induction is often sufficient to stop further pathogen growth/spread. Each plant genome contains a few hundred apparent R genes. Studied R genes usually confer specificity for particular pathogen strains. As first noted by Harold Flor in his mid-20th century formulation of the gene-for-gene relationship, the plant R gene and the pathogen Avr gene must have matched specificity for that R gene to confer resistance, suggesting a receptor/ligand interaction for Avr and R genes. Alternatively, an effector can modify its host cellular target (or a molecular decoy of that target) activating an NLR associated with the

Effector Biology

So-called "core" effectors are defined operationally by their wide distribution across the population of a particular pathogen and their substantial contribution to pathogen virulence. Genomics can be used to identify core effectors, which can then functionally define new R alleles, which can serve as breeding targets.

RNA Silencing and Systemic Acquired Resistance Elicited by Prior Infections

Against viruses, plants often induce pathogen-specific gene silencing mechanisms mediated by RNA interference.

Plant immune systems also can respond to an initial infection in one part of the plant by physiologically elevating the capacity for a successful defense response in other parts. Such responses include systemic acquired resistance, largely mediated by salicylic acid-dependent pathways, and induced systemic resistance, largely mediated.

Species-level Resistance

In a small number of cases, plant genes are effective against an entire pathogen species, even though that species that is pathogenic on other genotypes of that host species. Examples include barley MLO against powdery mildew, wheat Lr34 against leaf rust and wheat Yr36 against stripe rust. An array of mechanisms for this type of resistance may exist depending on the particular gene and plant-pathogen combination. Other reasons for effective plant immunity can include a lack of coadaptation (the pathogen and/or plant lack multiple mechanisms needed for colonization and growth within that host species), or a particularly effective suite of pre-formed defenses.

Signaling Mechanisms

Perception of Pathogen Presence

Plant defense signaling is activated by pathogen-detecting receptors. The activated receptors frequently elicit reactive oxygen and nitric oxide production, calcium, potassium and proton ion fluxes, altered levels of salicylic acid and other hormones and activation of MAP kinases and other specific protein kinases. These events in turn typically lead to the modification of proteins that control gene transcription, and the activation of defense-associated gene expression.

In addition to PTI and ETI, plant defenses can be activated by the sensing of damage-associated compounds (DAMP), such as portions of the plant cell wall released during pathogenic infection. Many receptors for MAMPs, effectors and DAMPs have been discovered. Effectors are often detected by NLRs, while MAMPs and DAMPs are often detected by transmembrane receptor-kinases that carry LRR or LysM extracellular domains.

Transcription Factors and The Hormone Response

Numerous genes and/or proteins as well as other molecules have been identified that mediate plant defense signal transduction. Cytoskeleton and vesicle trafficking dynamics help to orient plant defense responses toward the point of pathogen attack.

Mechanisms of Transcription Factors and Hormones

Plant immune system activity is regulated in part by signaling hormones such as:

- Salicylic acid

- Jasmonic acid

- Ethylene

There can be substantial cross-talk among these pathways.

Regulation by Degradation

As with many signal transduction pathways, plant gene expression during immune responses can be regulated by degradation. This often occurs when hormone binding to hormone receptors stimulates ubiquitin-associated degradation of repressor proteins that block expression of certain genes. The net result is hormone-activated gene expression. Examples:

- Auxin: binds to receptors that then recruit and degrade repressors of transcriptional activators that stimulate auxin-specific gene expression.

- Jasmonic acid: similar to auxin, except with jasmonate receptors impacting jasmonate-response signaling mediators such as JAZ proteins.

- Gibberellic acid: Gibberellin causes receptor conformational changes and binding and degradation of Della proteins.

- Ethylene: Inhibitory phosphorylation of the EIN2 ethylene response activator is blocked by ethylene binding. When this phosphorylation is reduced, EIN2 protein is cleaved and a portion of the protein moves to the nucleus to activate ethylene-response gene expression.

Ubiquitin and E3 Signaling

Ubiquitination plays a central role in cell signaling that regulates processes including protein degradation and immunological response. Although one of the main functions of ubiquitin is to target proteins for destruction, it is also useful in signaling pathways, hormone release, apoptosis and translocation of materials throughout the cell. Ubiquitination is a component of several immune responses. Without ubiquitin's proper functioning, the invasion of pathogens and other harmful molecules would increase dramatically due to weakened immune defenses.

E3 Signaling

The E3 Ubiquitin ligase enzyme is a main component that provides specificity in protein degradation pathways, including immune signaling pathways. The E3 enzyme components can be grouped by which domains they contain and include several types. These include the Ring and U-box single subunit, HECT, and CRLs. Plant signaling pathways including immune responses are controlled by several feedback pathways, which often include negative feedback; and they can be regulated by De-ubiquitination enzymes, degradation of transcription factors and the degradation of negative regulators of transcription.

Plant Breeding for Disease Resistance

Plant breeders emphasize selection and development of disease-resistant plant lines. Plant diseases can also be partially controlled by use of pesticides and by cultivation practices such as crop rotation, tillage, planting density, disease-free seeds and cleaning of equipment, but plant varieties with inherent (genetically determined) disease resistance are generally preferred. Breeding for disease resistance began when plants were first domesticated. Breeding efforts continue because pathogen populations are under selection pressure for increased virulence, new pathogens appear, evolving cultivation practices and changing climate can reduce resistance and/or strengthen

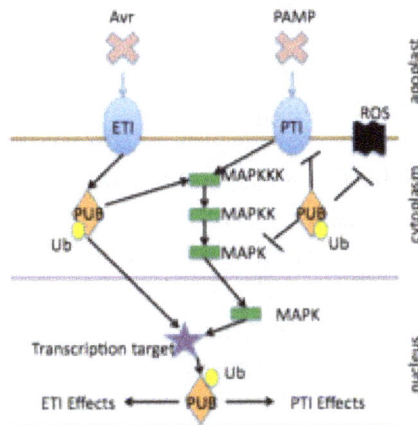

This image depicts the pathways taken during responses in plant immunity. It highlights the role and effect ubiquitin has in regulating the pathway.

pathogens, and plant breeding for other traits can disrupt prior resistance. A plant line with acceptable resistance against one pathogen may lack resistance against others.

Breeding for resistance typically includes:

- Identification of plants that may be less desirable in other ways, but which carry a useful disease resistance trait, including wild strains that often express enhanced resistance.

- Crossing of a desirable but disease-susceptible variety to another variety that is a source of resistance.

- Growth of breeding candidates in a disease-conducive setting, possibly including pathogen inoculation. Attention must be paid to the specific pathogen isolates, to address variability within a single pathogen species.

- Selection of disease-resistant individuals that retain other desirable traits such as yield, quality and including other disease resistance traits.

Resistance is termed *durable* if it continues to be effective over multiple years of widespread use as pathogen populations evolve. "Vertical resistance" is specific to certain races or strains of a pathogen species, is often controlled by single R genes and can be less durable. Hoizontal or broad-spectrum resistance against an entire pathogen species is often only incompletely effective, but more durable, and is often controlled by many genes that segregate in breeding populations.

Crops such as potato, apple, banana and sugarcane are often propagated by vegetative reproduction to preserve highly desirable plant varieties, because for these species, outcrossing seriously disrupts the preferred traits. Vegetatively propagated crops may be among the best targets for resistance improvement by the biotechnology method of plant transformation to manage genes that affect disease resistance.

Scientific breeding for disease resistance originated with Sir Rowland Biffen, who identified a single recessive gene for resistance to wheat yellow rust. Nearly every crop was then bred to include disease resistance (R) genes, many by introgression from compatible wild relatives.

GM or Transgenic Engineered Disease Resistance

The term GM ("genetically modified") is often used as a synonym of transgenic to refer to plants modified using recombinant DNA technologies. Plants with transgenic/GM disease resistance against insect pests have been extremely successful as commercial products, especially in maize and cotton, and are planted annually on over 20 million hectares in over 20 countries worldwide. Transgenic plant disease resistance against microbial pathogens was first demonstrated in 1986. Expression of viral coat protein gene sequences conferred virus resistance via small RNAs. This proved to be a widely applicable mechanism for inhibiting viral replication. Combining coat protein genes from three different viruses, scientists developed squash hybrids with field-validated, multiviral resistance. Similar levels of resistance to this variety of viruses had not been achieved by conventional breeding.

A similar strategy was deployed to combat papaya ringspot virus, which by 1994 threatened to destroy Hawaii's papaya industry. Field trials demonstrated excellent efficacy and high fruit quality.

By 1998 the first transgenic virus-resistant papaya was approved for sale. Disease resistance has been durable for over 15 years. Transgenic papaya accounts for ~85% of Hawaiian production. The fruit is approved for sale in the U.S., Canada and Japan.

Potato lines expressing viral replicase sequences that confer resistance to potato leafroll virus were sold under the trade names NewLeaf Y and NewLeaf Plus, and were widely accepted in commercial production in 1999-2001, until McDonald's Corp. decided not to purchase GM potatoes and Monsanto decided to close their NatureMark potato business. NewLeaf Y and NewLeaf Plus potatoes carried two GM traits, as they also expressed Bt-mediated resistance to Colorado potato beetle.

No other crop with engineered disease resistance against microbial pathogens had reached the market by 2013, although more than a dozen were in some state of development and testing.

Examples of transgenic disease resistance projects				
Publication year	Crop	Disease resistance	Mechanism	Development status
2012	Tomato	Bacterial spot	R gene from pepper	8 years of field trials
2012	Rice	Bacterial blight and bacterial streak	Engineered E gene	Laboratory
2012	Wheat	Powdery mildew	Overexpressed R gene from wheat	2 years of field trials at time of publication
2011	Apple	Apple scab fungus	Thionin gene from barley	4 years of field trials at time of publication
2011	Potato	Potato virus Y	Pathogen-derived resistance	1 year of field trial at time of publication
2010	Apple	Fire blight	Antibacterial protein from moth	12 years of field trials at time of publication
2010	Tomato	Multibacterial resistance	PRR from *Arabidopsis*	Laboratory scale
2010	Banana	Xanthomonas wilt	Novel gene from pepper	Now in field trial
2009	Potato	Late blight	R genes from wild relatives	3 years of field trials
2009	Potato	Late blight	R gene from wild relative	2 years of field trials at time of publication
2008	Potato	Late blight	R gene from wild relative	2 years of field trials at time of publication
2008	Plum	Plum pox virus	Pathogen-derived resistance	Regulatory approvals, no commercial sales
2005	Rice	Bacterial streak	R gene from maize	Laboratory
2002	Barley	Stem rust	Resting lymphocyte kinase (RLK) gene from resistant barley cultivar	Laboratory
1997	Papaya	Ring spot virus	Pathogen-derived resistance	Approved and commercially sold since 1998, sold into Japan since 2012
1995	Squash	Three mosaic viruses	Pathogen-derived resistance	Approved and commercially sold since 1994
1993	Potato	Potato virus X	Mammalian interferon-induced enzyme	3 years of field trials at time of publication

PRR Transfer

Research aimed at engineered resistance follows multiple strategies. One is to transfer useful PRRs into species that lack them. Identification of functional PRRs and their transfer to a recipient species that lacks an orthologous receptor could provide a general pathway to additional broadened PRR repertoires. For example, the *Arabidopsis* PRR *EF-Tu* receptor (EFR) recognizes the bacterial translation elongation factor *EF-Tu*. Research performed at Sainsbury Laboratory demonstrated that deployment of EFR into either *Nicotiana benthamiana* or *Solanum lycopersicum* (tomato), which cannot recognize *EF-Tu*, conferred resistance to a wide range of bacterial pathogens. EFR expression in tomato was especially effective against the widespread and devastating soil bacterium Ralstonia solanacearum. Conversely, the tomato PRR *Verticillium 1* (*Ve1*) gene can be transferred from tomato to *Arabidopsis*, where it confers resistance to race 1 Verticillium isolates.

Stacking

The second strategy attempts to deploy multiple NLR genes simultaneously, a breeding strategy known as stacking. Cultivars generated by either DNA-assisted molecular breeding or gene transfer will likely display more durable resistance, because pathogens would have to mutate multiple effector genes. DNA sequencing allows researchers to functionally "mine" NLR genes from multiple species/strains.

The *avrBs2* effector gene from *Xanthomona perforans* is the causal agent of bacterial spot disease of pepper and tomato. The first "effector-rationalized" search for a potentially durable R gene followed the finding that *avrBs2* is found in most disease-causing *Xanthomonas* species and is required for pathogen fitness. The *Bs2* NLR gene from the wild pepper, *Capsicum chacoense*, was moved into tomato, where it inhibited pathogen growth. Field trials demonstrated robust resistance without bactericidal chemicals. However, rare strains of *Xanthomonas* overcame *Bs2*-mediated resistance in pepper by acquisition of *avrBs2* mutations that avoid recognition but retain virulence. Stacking R genes that each recognize a different core effector could delay or prevent adaptation.

More than 50 loci in wheat strains confer disease resistance against wheat stem, leaf and yellow stripe rust pathogens. The Stem rust 35 (*Sr35*) NLR gene, cloned from a diploid relative of cultivated wheat, *Triticum monococcum*, provides resistance to wheat rust isolate *Ug99*. Similarly, *Sr33*, from the wheat relative *Aegilops tauschii*, encodes a wheat ortholog to barley *Mla* powdery mildew–resistance genes. Both genes are unusual in wheat and its relatives. Combined with the *Sr2* gene that acts additively with at least Sr33, they could provide durable disease resistance to *Ug99* and its derivatives.

Executor Genes

Another class of plant disease resistance genes opens a "trap door" that quickly kills invaded cells, stopping pathogen proliferation. Xanthomonas and Ralstonia transcription activator–like (TAL) effectors are DNA-binding proteins that activate host gene expression to enhance pathogen virulence. Both the rice and pepper lineages independently evolved TAL-effector binding sites that instead act as an executioner that induces hypersensitive host cell death when up-regulated. *Xa27* from rice and Bs3 and Bs4c from pepper, are such "executor" (or "executioner") genes that encode non-homologous plant proteins of unknown function. Executor genes are expressed only in the presence of a specific TAL effector.

Engineered executor genes were demonstrated by successfully redesigning the pepper *Bs3* promoter to contain two additional binding sites for TAL effectors from disparate pathogen strains. Subsequently, an engineered executor gene was deployed in rice by adding five TAL effector binding sites to the *Xa27* promoter. The synthetic *Xa27* construct conferred resistance against Xanthomonas bacterial blight and bacterial leaf streak species.

Host Susceptibility Alleles

Most plant pathogens reprogram host gene expression patterns to directly benefit the pathogen. Reprogrammed genes required for pathogen survival and proliferation can be thought of as "disease-susceptibility genes." Recessive resistance genes are disease-susceptibility candidates. For example, a mutation disabled an *Arabidopsis* gene encoding pectate lyase (involved in cell wall degradation), conferring resistance to the powdery mildew pathogen *Golovinomyces cichoracearum*. Similarly, the Barley *MLO* gene and spontaneously mutated pea and tomato *MLO* orthologs also confer powdery mildew resistance.

Lr34 is a gene that provides partial resistance to leaf and yellow rusts and powdery mildew in wheat. *Lr34* encodes an adenosine triphosphate (ATP)–binding cassette (ABC) transporter. The dominant allele that provides disease resistance was recently found in cultivated wheat (not in wild strains) and, like *MLO* provides broad-spectrum resistance in barley.

Natural alleles of host translation elongation initiation factors *eif4e* and *eif4g* are also recessive viral-resistance genes. Some have been deployed to control potyviruses in barley, rice, tomato, pepper, pea, lettuce and melon. The discovery prompted a successful mutant screen for chemically induced *eif4e* alleles in tomato.

Natural promoter variation can lead to the evolution of recessive disease-resistance alleles. For example, the recessive resistance gene *xa13* in rice is an allele of *Os-8N3*. *Os-8N3* is transcriptionally activated by*Xanthomonas oryzae pv. oryzae* strains that express the TAL effector *PthXo1*. The *xa13* gene has a mutated effector-binding element in its promoter that eliminates *PthXo1* binding and renders these lines resistant to strains that rely on *PthXo1*. This finding also demonstrated that *Os-8N3* is required for susceptibility.

Xa13/Os-8N3 is required for pollen development, showing that such mutant alleles can be problematic should the disease-susceptibility phenotype alter function in other processes. However, mutations in the *Os11N3* (OsSWEET14) TAL effector–binding element were made by fusing TAL effectors to nucleases (TALENs). Genome-edited rice plants with altered *Os11N3* binding sites remained resistant to *Xanthomonas oryzae pv. oryzae*, but still provided normal development function.

Gene Silencing

RNA silencing-based resistance is a powerful tool for engineering resistant crops. The advantage of RNAi as a novel gene therapy against fungal, viral and bacterial infection in plants lies in the fact that it regulates gene expression via messenger RNA degradation, translation repression and chromatin remodelling through small non-coding RNAs. Mechanistically, the silencing processes are guided by processing products of the double-stranded RNA (dsRNA) trigger, which are known as small interfering RNAs and microRNAs.

Host Range

Among the thousands of species of plant pathogenic microorganisms, only a small minority have the capacity to infect a broad range of plant species. Most pathogens instead exhibit a high degree of host-specificity. Non-host plant species are often said to express *non-host resistance*. The term *host resistance* is used when a pathogen species can be pathogenic on the host species but certain strains of that plant species resist certain strains of the pathogen species. The causes of host resistance and non-host resistance can overlap. Pathogen host range can change quite suddenly if, for example, the pathogen's capacity to synthesize a host-specific toxin or effector is gained by gene shuffling/mutation, or by horizontal gene transfer.

Epidemics and Population Biology

Native populations are often characterized by substantial genotype diversity and dispersed populations (growth in a mixture with many other plant species). They also have undergone of plant-pathogen coevolution. Hence as long as novel pathogens are not introduced/do not evolve, such populations generally exhibit only a low incidence of severe disease epidemics.

Monocrop agricultural systems provide an ideal environment for pathogen evolution, because they offer a high density of target specimens with similar/identical genotypes.

The rise in mobility stemming from modern transportation systems provides pathogens with access to more potential targets.

Climate change can alter the viable geographic range of pathogen species and cause some diseases to become a problem in areas where the disease was previously less important.

These factors make modern agriculture more prone to disease epidemics. Common solutions include constant breeding for disease resistance, use of pesticides, use of border inspections and plant import restrictions, maintenance of significant genetic diversity within the crop gene pool, and constant surveillance to accelerate initiation of appropriate responses. Some pathogen species have much greater capacity to overcome plant disease resistance than others, often because of their ability to evolve rapidly and to disperse broadly.

Pest Control

Pest control refers to the regulation or management of a species defined as a pest,and can be perceived to be detrimental to a person's health, the ecology or the economy. A practitioner of pest control is called an exterminator.

History

Pest control is at least as old as agriculture, as there has always been a need to keep crops free from pests. In order to maximize food production, it is advantageous to protect crops from competing species of plants, as well as from herbivores competing with humans.

The conventional approach was probably the first to be employed, since it is comparatively easy to destroy weeds by burning them or plowing them under, and to kill larger competing herbivores, such as crows and other birds eating seeds. Techniques such as crop rotation, companion planting (also known as intercropping or mixed cropping), and the selective breeding of pest-resistant cultivars have a long history.

In the UK, following concern about animal welfare, humane pest control and deterrence is gaining ground through the use of animal psychology rather than destruction. For instance, with the urban red fox which territorial behaviour is used against the animal, usually in conjunction with non-injurious chemical repellents. In rural areas of Britain, the use of firearms for pest control is quite common. Airguns are particularly popular for control of small pests such as rats, rabbits and grey squirrels, because of their lower power they can be used in more restrictive spaces such as gardens, where using a firearm would be unsafe.

Chemical pesticides date back 4,500 years, when the Sumerians used sulfur compounds as insecticides. The Rig Veda, which is about 4,000 years old, also mentions the use of poisonous plants for pest control. It was only with the industrialization and mechanization of agriculture in the 18th and 19th century, and the introduction of the insecticides pyrethrum and derris that chemical pest control became widespread. In the 20th century, the discovery of several synthetic insecticides, such as DDT, and herbicides boosted this development. Chemical pest control is still the predominant type of pest control today, although its long-term effects led to a renewed interest in traditional and biological pest control towards the end of the 20th century.

Causes

Many pests have only become a problem as a result of the direct actions by humans. Modifying these actions can often substantially reduce the pest problem. In the United States, raccoons caused a nuisance by tearing open refuse sacks. Many householders introduced bins with locking lids, which deterred the raccoons from visiting. House flies tend to accumulate wherever there is human activity and is virtually a global phenomenon, especially where food or food waste is exposed. Similarly, seagulls have become pests at many seaside resorts. Tourists would often feed the birds with scraps of fish and chips, and before long, the birds would rely on this food source and act aggressively towards humans.

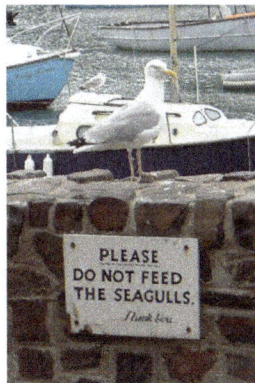

Sign in Ilfracombe, England designed to help control seagull presence

Living organisms evolve and increase their resistance to biological, chemical, physical or any other form of control. Unless the target population is completely exterminated or is rendered incapable of reproduction, the surviving population will inevitably acquire a tolerance of whatever pressures are brought to bear - this results in an evolutionary arms race.

Types of Pest Control

Use of Pest-destroying Animals

Perhaps as far ago as 3000BC in Egypt, cats were being used to control pests of grain stores such as rodents. In 1939/40 a survey discovered that cats could keep a farm's population of rats down to a low level, but could not eliminate them completely. However, if the rats were cleared by trapping or poisoning, farm cats could stop them returning - at least from an area of 50 yards around a barn.

Ferrets were domesticated at least by 500 AD in Europe, being used as mousers. Mongooses have been introduced into homes to control rodents and snakes, probably at first by the ancient Egyptians.

Biological Pest Control

Biological pest control is the control of one through the control and management of natural predators and parasites. For example: mosquitoes are often controlled by putting *Bt Bacillus thuringiensis* ssp. *israelensis*, a bacterium that infects and kills mosquito larvae, in local water sources. The treatment has no known negative consequences on the remaining ecology and is safe for humans to drink. The point of biological pest control, or any natural pest control, is to eliminate a pest with minimal harm to the ecological balance of the environment in its present form.

Mechanical Pest Control

Mechanical pest control is the use of hands-on techniques as well as simple equipment and devices, that provides a protective barrier between plants and insects. For example: weeds can be controlled by being physically removed from the ground. This is referred to as tillage and is one of the oldest methods of weed control.

Physical Pest Control

Dog control van, Rekong Peo, Himachal Pradesh, India

Physical pest control is a method of getting rid of insects and small rodents by removing, attacking, setting up barriers that will prevent further destruction of one's plants, or forcing insect infestations to become visual.

Elimination of Breeding Grounds

Proper waste management and drainage of still water, eliminates the breeding ground of many pests.

Garbage provides food and shelter for many unwanted organisms, as well as an area where still water might collect and be used as a breeding ground by mosquitoes. Communities that have proper garbage collection and disposal, have far less of a problem with rats, cockroaches, mosquitoes, flies and other pests than those that don't.

Open air sewers are ample breeding ground for various pests as well. By building and maintaining a proper sewer system, this problem is eliminated.

Certain spectrums of LED light can "disrupt insects' breeding".

Poisoned Bait

Poisoned bait is a common method for controlling rat populations, however is not as effective when there are other food sources around, such as garbage. Poisoned meats have been used for centuries for killing off wolves, birds that were seen to threaten crops, and against other creatures. This can be a problem, since a carcass which has been poisoned will kill not only the targeted animal, but also every other animal which feeds on the carcass. Humans have also been killed by coming in contact with poisoned meat, or by eating an animal which had fed on a poisoned carcass. This tool is also used to manage several caterpillars e.g. Spodoptera litura, fruit flies, snails and slugs, crabs etc.

Field Burning

Traditionally, after a sugar cane harvest, the fields are all burned, to kill off any rodents, insects or eggs that might be in the fields.

Hunting

Historically, in some European countries, when stray dogs and cats became too numerous, local populations gathered together to round up all animals that did not appear to have an owner and kill them. In some nations, teams of rat-catchers work at chasing rats from the field, and killing them with dogs and simple hand tools. Some communities have in the past employed a bounty system, where a town clerk will pay a set fee for every rat head brought in as proof of a rat killing.

Traps

A variety of mouse traps and rat traps are available for mice and rats, including snap traps, glue traps and live catch traps.

Pesticides

Spraying pesticides by planes, trucks or by hand is a common method of pest control. Crop dusters commonly fly over farmland and spray pesticides to kill off pests that would threaten the crops. However, some pesticides may cause cancer and other health problems, as well as harming wildlife.

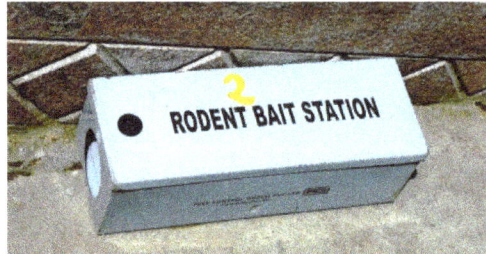

Rodent bait station, Chennai, India

Space Fumigation

A project that involves a structure be covered or sealed airtight followed by the introduction of a penetrating, deadly gas at a killing concentration a long period of time (24-72hrs.). Although expensive, space fumigation targets all life stages of pests.

Space Treatment

A long term project involving fogging or misting type applicators. Liquid insecticide is dispersed in the atmosphere within a structure. Treatments do not require the evacuation or airtight sealing of a building, allowing most work within the building to continue but at the cost of the penetrating effects. Contact insecticides are generally used, minimizing the long lasting residual effects. On August 10, 1973, the Federal Register printed the definition of Space treatment as defined by the U.S. Environmental Protection Agency (EPA):

Residential & commercial building pest control service vehicle, Ypsilanti Township, Michigan

the dispersal of insecticides into the air by foggers, misters, aerosol devices or vapor dispensers for control of flying insects and exposed crawling insects

Sterilization

Laboratory studies conducted with U-5897 (3-chloro-1,2-propanediol) were attempted in the early 1970s although these proved unsuccessful. Research into sterilization bait is ongoing.

In 2013, New York City tested sterilization traps in a $1.1 million study. The result was a 43% reduction in rat populations. The Chicago Transit Authority plans to test sterilization control in spring 2015. The sterilization method doesn't poison the rats or humans.

Destruction of Infected Plants

Forest services sometimes destroy all the trees in an area where some are infected with insects, if seen as necessary to prevent the insect species from spreading. Farms infested with certain insects, have been burned entirely, to prevent the pest from spreading elsewhere.

Natural Rodent Control

Several wildlife rehabilitation organizations encourage natural form of rodent control through exclusion and predator support and preventing secondary poisoning altogether.

Example of House mouse infestation

The United States Environmental Protection Agency agrees, noting in its Proposed Risk Mitigation Decision for Nine Rodenticides that "without habitat modification to make areas less attractive to commensal rodents, even eradication will not prevent new populations from recolonizing the habitat."

Repellents

- Balsam fir oil from the tree *Abies balsamea* is an EPA approved non-toxic rodent repellent.

- *Acacia polyacantha* subsp. *campylacantha* root emits chemical compounds that repel animals including crocodiles, snakes and rats.

Types of Pest Control

Biological Pest Control

Biological control is a method of controlling pests such as insects, mites, weeds and plant diseases using other organisms. It relies on predation, parasitism, herbivory, or other natural mechanisms, but typically also involves an active human management role. It can be an important component of integrated pest management (IPM) programs.

Syrphus hoverfly larva feeding on aphids

There are three basic types of biological pest control strategies: importation (sometimes called classical biological control), in which a natural enemy of a pest is introduced in the hope of achieving control; augmentation, in which locally-occurring natural enemies are bred and released to improve control; and conservation, in which measures are taken to increase natural enemies, such as by planting nectar-producing crop plants in the borders of rice fields.

Parasitic wasp *Cotesia congregata* on tobacco hornworm *Manduca sexta*

Natural enemies of insect pests, also known as biological control agents, include predators, parasitoids, and pathogens. Biological control agents of plant diseases are most often referred to as antagonists. Biological control agents of weeds include seed predators, herbivores and plant pathogens.

Biological control can have side-effects on biodiversity through predation, parasitism, pathogenicity, competition, or other attacks on non-target species, especially when a species is introduced without thorough understanding of the possible consequences.

History

The term "biological control" was first used by Harry Scott Smith at the 1919 meeting of the Pacific Slope Branch of the American Association of Economic Entomologists, at the Mission Inn in downtown Riverside, California, and later defined by P. DeBach and K. S. Hagen in 1964. However, the practice has previously been used for centuries. The first report of the use of an insect species to control an insect pest comes from "Nan Fang Cao Mu Zhuang" (南方草木狀 *Plants of the Southern Regions*) (ca. 304 AD), which is attributed to Western Jin dynasty botanist *Ji Han* (嵇含, 263-307), in which it is mentioned that "*Jiaozhi people sell ants and their nests attached to twigs looking like thin cotton envelopes, the reddish-yellow ant being larger than normal. Without such ants, southern citrus fruits will be severely insect-damaged*". The ants used are known as *huang gan* (*huang* = yellow, *gan* = citrus) ants (*Oecophylla smaragdina*). The practice was later reported by Ling Biao Lu Yi (late Tang Dynasty or Early Five Dynasties), in *Ji Le Pian* by *Zhuang Jisu* (Southern Song Dynasty), in the *Book of Tree Planting* by Yu Zhen Mu (Ming Dynasty), in the book *Guangdong Xing Yu* (17th century), *Lingnan* by Wu Zhen Fang (Qing Dynasty), in *Nanyue Miscellanies* by Li Diao Yuan, and others.

Biological control techniques as we know them today started to emerge in the 1870s. During this decade, in the USA, the Missouri State Entomologist C. V. Riley and the Illinois State Entomologist W. LeBaron began within-state redistribution of parasitoids to control crop pests. The first international shipment of an insect as biological control agent was made by Charles V. Riley in 1873, shipping to France the predatory mites *Tyroglyphus phylloxera* to help fight the grapevine phylloxera (*Daktulosphaira vitifoliae*) that was destroying grapevines in France. The United States Department of Agriculture (USDA) initiated research in classical biological control following the establishment of the Division of Entomology in 1881, with C. V. Riley as Chief. The first importation of a parasitoid into the United States was this of *Cotesia glomerata* in 1883-1884, imported from Europe to control the imported cabbage white butterfly, *Pieris rapae*. In 1888-1889 the vedalia beetle, *Rodolia cardinalis*, which is a ladybug, was introduced from Australia to California to control the cottony cushion scale, *Icerya purchasi*. This had become a major problem for the newly developed citrus industry in California, and by the end of 1889 the cottony cushion scale population had already declined. This great success led to further introductions of beneficial insects into the USA.

In 1905 the USDA initiated its first large-scale biological control program, sending entomologists to Europe and Japan to look for natural enemies of the gypsy moth, *Lymantria dispar dispar*, and brown-tail moth, *Euproctis chrysorrhoea*, invasive pests of trees and shrubs. As a result, nine parasitoids of gypsy moth, seven of brown-tail moth, and two predators for both moths became established in the USA. Although the gypsy moth was not fully controlled by these natural enemies, the frequency, duration, and severity of its outbreaks were reduced and the program was regarded as successful. This program also led to the development of many concepts, principles, and procedures for the implementation of biological control programs.

The first reported case of a classical biological control attempt in Canada involves the hymenopteran parasitoid *Trichogramma minutum*. Individuals were caught in New York State and released in Ontario gardens in 1882 by William Saunders, trained chemist and first Director of the Dominion Experimental Farms, for controlling the imported currantworm *Nematus ribesii*. Between 1884 and 1908, the first Dominion Entomologist, James Fletcher, continued introductions of other parasitoids and pathogens for the control of pests in Canada.

Types of Biological Pest Control

There are three basic biological pest control strategies: importation (classical biological control), augmentation and conservation.

Importation

Importation or classical biological control involves the introduction of a pest's natural enemies to a new locale where they do not occur naturally. Early instances were often unofficial and not based on research, and some introduced species became serious pests themselves.

Rodolia cardinalis, the vedalia beetle, was imported to Australia in the 19th century, successfully controlling cottony cushion scale.

To be most effective at controlling a pest, a biological control agent requires a colonizing ability which allows it to keep pace with the spatial and temporal disruption of the habitat. Control is greatest if the agent has temporal persistence, so that it can maintain its population even in the temporary absence of the target species, and if it is an opportunistic forager, enabling it to rapidly exploit a pest population.

Joseph Needham noted a Chinese text dating from 304 AD, *Records of the Plants and Trees of the Southern Regions*, by Hsi Han, which describes mandarin oranges protected by large reddish-yellow citrus ants which attack and kill insect pests of the orange trees. The citrus ant (*Oecophylla smaragdina*) was rediscovered in the 20th century, and since 1958 has been used in China to protect orange groves.

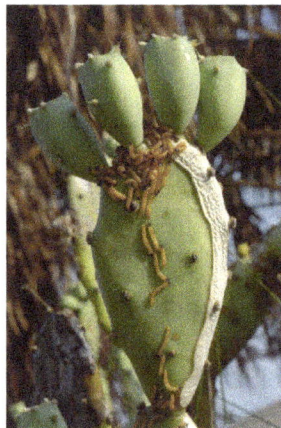

Cactoblastis cactorum larvae feeding on *Opuntia* cacti

One of the earliest successes in the west was in controlling *Icerya purchasi* (cottony cushion scale) in Australia, using a predatory insect *Rodolia cardinalis* (the vedalia beetle). This success was repeated in California using the beetle and a parasitoid fly, *Cryptochaetum iceryae*.

Prickly pear cacti were introduced into Queensland, Australia as ornamental plants. They quickly spread to cover over 25 million hectares of Australia. Two control agents were used to help control the spread of the plant, the cactus moth *Cactoblastis cactorum*, and *Dactylopius* scale insects.

Damage from *Hypera postica*, the alfalfa weevil, a serious introduced pest of forage, was substantially reduced by the introduction of natural enemies. 20 years after their introduction the population of weevils in the alfalfa area treated for alfalfa weevil in the Northeastern United States remained 75 percent down.

The invasive species *Alternanthera philoxeroides* (alligator weed) was controlled in Florida (U.S.) by introducing alligator weed flea beetle.

Alligator weed was introduced to the United States from South America. It takes root in shallow water, interfering with navigation, irrigation, and flood control. The alligator weed flea beetle and two other biological controls were released in Florida, enabling the state to ban the use of herbicides to control alligator weed three years later. Another aquatic weed, the giant salvinia (*Salvinia molesta*) is a serious pest, covering waterways, reducing water flow and harming native species. Control with the salvinia weevil (*Cyrtobagous salviniae*) is effective in warm climates, and in Zimbabwe, a 99% control of the weed was obtained over a two-year period.

Small commercially reared parasitoidal wasps, *Trichogramma ostriniae*, provide limited and erratic control of the European corn borer (*Ostrinia nubilalis*), a serious pest. Careful formulations of the bacterium *Bacillus thuringiensis* are more effective.

The population of *Levuana iridescens*, the Levuana moth, a serious coconut pest in Fiji, was brought under control by a classical biological control program in the 1920s.

Augmentation

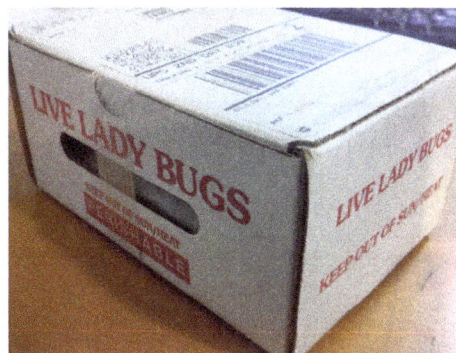

Hippodamia convergens, the convergent lady beetle, is commonly sold for biological control of aphids.

Augmentation involves the supplemental release of natural enemies, boosting the naturally occurring population. In inoculative release, small numbers of the control agents are released at intervals to allow them to reproduce, in the hope of setting up longer-term control, and thus keeping the pest down to a low level, constituting prevention rather than cure. In inundative release, in contrast, large numbers are released in the hope of rapidly reducing a damaging pest population, correcting a problem that has already arisen. Augmentation can be effective, but is not guaranteed to work, and relies on understanding of the situation.

An example of inoculative release occurs in greenhouse production of several crops. Periodic releases of the parasitoid, *Encarsia formosa*, are used to control greenhouse whitefly, while the predatory mite *Phytoseiulus persimilis* is used for control of the two-spotted spider mite.

The egg parasite *Trichogramma* is frequently released inundatively to control harmful moths. Similarly, *Bacillus thuringiensis* and other microbial insecticides are similarly used in large enough quantities for a rapid effect. Recommended release rates for *Trichogramma* in vegetable or field crops range from 5,000 to 200,000 per acre (1 to 50 per square metre) per week according to the level of pest infestation. Similarly, entomopathogenic nematodes are released at rates of millions and even billions per acre for control of certain soil-dwelling insect pests.

Conservation

The conservation of existing natural enemies in an environment is the third method of biological pest control. Natural enemies are already adapted to the habitat and to the target pest, and their conservation can be simple and cost-effective, as when nectar-producing crop plants are grown in the borders of rice fields. These provide nectar to support parasitoids and predators of planthopper pests and have been demonstrated to be so effective (reducing pest densities by 10- or even 100-fold) that farmers sprayed 70% less insecticides, enjoyed yields boosted by 5%, and this led to an economic advantage of 7.5%. Predators of aphids were similarly found to be present in tussock grasses by field boundary hedges in England, but they spread too slowly to reach the centres of fields. Control was improved by planting a metre-wide strip of tussock grasses in field centres, enabling aphid predators to overwinter there.

An inverted flowerpot filled with straw to attract earwigs

Cropping systems can be modified to favor natural enemies, a practice sometimes referred to as habitat manipulation. Providing a suitable habitat, such as a shelterbelt, hedgerow, or beetle bank where beneficial insects can live and reproduce, can help ensure the survival of populations of natural enemies. Things as simple as leaving a layer of fallen leaves or mulch in place provides a suitable food source for worms and provides a shelter for insects, in turn being a food source for such beneficial mammals as hedgehogs and shrews. Compost piles and stacks of wood can provide shelter for invertebrates and small mammals. Long grass and ponds support amphibians. Not removing dead annuals and non-hardy plants in the autumn allows insects to make use of their hollow stems during winter. In California, prune trees are sometimes planted in grape vineyards to provide an improved overwintering habitat or refuge for a key grape pest parasitoid. The providing of artificial shelters in the form of wooden caskets, boxes or flowerpots is also sometimes undertaken, particularly in gardens, to make a cropped area more attractive to natural enemies. For example, earwigs are natural predators which can be encouraged in gardens by hanging upside-down flowerpots filled with straw or wood wool. Green lacewings can be encouraged by using plastic bottles with an open bottom and a roll of cardboard inside. Birdhouses enable insectivorous birds to nest; the most useful birds can be attracted by choosing an opening just large enough for the desired species.

Biological Control Agents

Predators

Predators are mainly free-living species that directly consume a large number of prey during their whole lifetime. Ladybugs, and in particular their larvae which are active between May and July in the northern hemisphere, are voracious predators of aphids, and also consume mites, scale insects and small caterpillars. The spotted lady beetle (*Coleomegilla maculata*) is also able to feed on the eggs and larvae of the Colorado potato beetle (*Leptinotarsa decemlineata*).

Lacewings are available from biocontrol dealers.

The larvae of many hoverfly species principally feed upon greenfly (aphids), one larva devouring up to 400 in its lifetime. Their effectiveness in commercial crops has not been studied.

Several species of entomopathogenic nematode are important predators of insect and other invertebrate pests. *Phasmarhabditis hermaphrodita* is a microscopic nematode that kills slugs. Its complex life cycle include a free-living, infective stage in the soil where it becomes associated with a pathogenic bacteria such as *Moraxella osloensis*. The nematode enters the slug through the posterior mantle region, thereafter feeding and reproducing inside, but it is the bacteria that kill the slug. The nematode is available commercially in Europe and is applied by watering onto moist soil.

Predatory *Polistes* wasp looking for bollworms or other caterpillars on a cotton plant

Species used to control spider mites include the predatory mites *Phytoseiulus persimilis*, *Neoseilus californicus,* and *Amblyseius cucumeris*, the predatory midge *Feltiella acarisuga*, and a ladybird *Stethorus punctillum*. The bug *Orius insidiosus* has been successfully used against the two-spotted spider mite and the western flower thrips (*Frankliniella occidentalis*).

Parasitoids

Parasitoids lay their eggs on or in the body of an insect host, which is then used as a food for developing larvae. The host is ultimately killed. Most insect parasitoids are wasps or flies, and may have a very narrow host range. The most important groups are the ichneumonid wasps, which prey mainly on caterpillars of butterflies and moths; braconid wasps, which attack caterpillars and a wide range of other insects including greenfly; chalcid wasps, which parasitize eggs and larvae of greenfly, whitefly, cabbage caterpillars, and scale insects; and tachinid flies, which parasitize a wide range of insects including caterpillars, adult and larval beetles, and true bugs.

Encarsia formosa was one of the first biological control agents developed.

Encarsia formosa is a small predatory chalcid wasp which is a parasitoid of whitefly, a sap-feeding insect which can cause wilting and black sooty moulds in glasshouse vegetable and ornamental crops. It is most effective when dealing with low level infestations, giving protection over a long period of time. The wasp lays its eggs in young whitefly 'scales', turning them black as the parasite larvae pupates. *Gonatocerus ashmeadi* (Hymenoptera: Mymaridae) has been introduced to control the glassy-winged sharpshooter *Homalodisca vitripennis* (Hemipterae: Cicadellidae) in French Polynesia and has successfully controlled ~95% of the pest density.

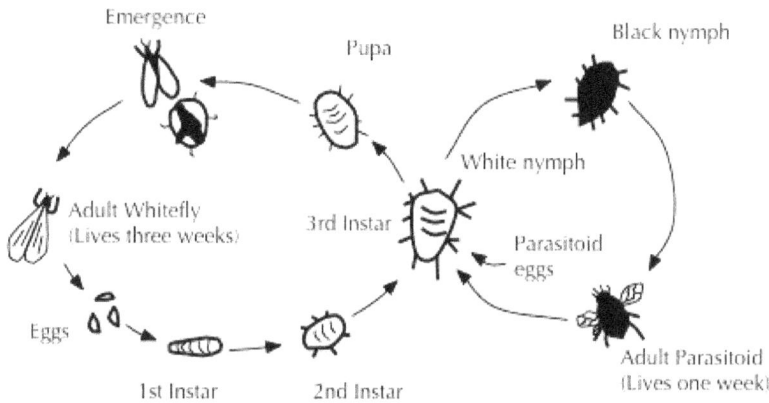

Life cycles of Greenhouse whitefly and its parasitoid wasp *Encarsia formosa*

Parasitoids are among the most widely used biological control agents. Commercially, there are two types of rearing systems: short-term daily output with high production of parasitoids per day, and long-term low daily output with a range in production of 4-1000million female parasitoids per week. Larger production facilities produce on a yearlong basis, whereas some facilities produce only seasonally. Rearing facilities are usually a significant distance from where the agents are to be used in the field, and transporting the parasitoids from the point of production to the point of use can pose problems. Shipping conditions can be too hot, and even vibrations from planes or trucks can adversely affect parasitoids.

Pathogens

Pathogenic micro-organisms include bacteria, fungi, and viruses. They kill or debilitate their host and are relatively host-specific. Various microbial insect diseases occur naturally, but may also be used as biological pesticides. When naturally occurring, these outbreaks are density-dependent in that they generally only occur as insect populations become denser.

Bacteria

Bacteria used for biological control infect insects via their digestive tracts, so they offer only limited options for controlling insects with sucking mouth parts such as aphids and scale insects. *Bacillus thuringiensis* is the most widely applied species of bacteria used for biological control, with at least four sub-species used against Lepidopteran (moth, butterfly), Coleopteran (beetle) and Dipteran (true fly) insect pests. The bacterium is available in sachets of dried spores which are mixed with water and sprayed onto vulnerable plants such as brassicas and fruit trees. *B. thuringiensis* has also been incorporated into crops, making them resistant to these pests and thus reducing the use of pesticides. The bacterium *Paenibacillus popilliae* causes milky spore disease has been found useful in the control of Japanese beetle, killing the larvae. It is very specific to its host species and is harmless to vertebrates and other invertebrates.

Fungi

Entomopathogenic fungi, which cause disease in insects, include at least 14 species that attack aphids. *Beauveria bassiana* is mass-produced and used to manage a wide variety of insect pests including whiteflies, thrips, aphids and weevils. *Lecanicillium* spp. are deployed against white flies, thrips and aphids. *Metarhizium* spp. are used against pests including beetles, locusts and other grasshoppers, Hemiptera, and spider mites. *Paecilomyces fumosoroseus* is effective against white flies, thrips and aphids; *Purpureocillium lilacinus* is used against root-knot nematodes, and 89 *Trichoderma* species against certain plant pathogens. *Trichoderma viride* has been used against Dutch elm disease, and has shown some effect in suppressing silver leaf, a disease of stone fruits caused by the pathogenic fungus *Chondrostereum purpureum*.

Green peach aphid, a pest in its own right and a vector of plant viruses, killed by the fungus *Pandora neoaphidis* (Zygomycota: Entomophthorales) Scale bar = 0.3 mm.

The fungi *Cordyceps* and *Metacordyceps* are deployed against a wide spectrum of arthropods. *Entomophaga* is effective against pests such as the green peach aphid.

Several members of Chytridiomycota and Blastocladiomycota have been explored as agents of biological control. From Chytridiomycota, *Synchytrium solstitiale* is being considered as a control agent of the yellow star thistle (*Centaurea solstitialis*) in the United States.

Viruses

Baculoviruses are specific to individual insect host species and have been shown to be useful in biological pest control. For example, the Lymantria dispar multicapsid nuclear polyhedrosis virus has been used to spray large areas of forest in North America where larvae of the gypsy moth are causing serious defoliation. The moth larvae are killed by the virus they have eaten and die, the disintegrating cadavers leaving virus particles on the foliage to infect other larvae.

A mammalian virus, the rabbit haemorrhagic disease virus has been introduced to Australia and to New Zealand to attempt to control the European rabbit populations there.

Algae

Lagenidium giganteum is a water-borne mould that parasitizes the larval stage of mosquitoes. When applied to water, the motile spores avoid unsuitable host species and search out suitable mosquito larval hosts. This alga has the advantages of a dormant phase, resistant to desiccation, with slow-release characteristics over several years. Unfortunately, it is susceptible to many chemicals used in mosquito abatement programmes.

Plants

The legume vine *Mucuna pruriens* is used in the countries of Benin and Vietnam as a biological control for problematic *Imperata cylindrica* grass. *Mucuna pruriens* is said not to be invasive outside its cultivated area. *Desmodium uncinatum* can be used in push-pull farming to stop the parasitic plant, *Striga*.

Other Methods

Combined Use of Parasitoids and Pathogens

In cases of massive and severe infection of invasive pests, techniques of pest control are often used in combination. An example is the emerald ash borer, *Agrilus planipennis*, an invasive beetle from China, which has destroyed tens of millions of ash trees in its introduced range in North America. As part of the campaign against it, from 2003 American scientists and the Chinese Academy of Forestry searched for its natural enemies in the wild, leading to the discovery of several parasitoid wasps, namely *Tetrastichus planipennisi*, a gregarious larval endoparasitoid,*Oobius agrili*, a solitary, parthenogenic egg parasitoid, and *Spathius agrili*, a gregarious larval ectoparasitoid. These have been introduced and released into the United States of America as a possible biological control of the emerald ash borer. Initial results have shown promise with *Tetrastichus planipennisi* and it is now being released along with *Beauveria bassiana*, a fungal pathogen with known insecticidal properties.

Indirect Control

Pests may be controlled by biological control agents that do not prey directly upon them. For example, the Australian bush fly, *Musca vetustissima*, is a major nuisance pest in Australia, but native decomposers found in Australia are not adapted to feeding on cow dung, which is where bush flies breed. Therefore, the Australian Dung Beetle Project (1965–1985), led by Dr. George Bornemissza of the Commonwealth Scientific and Industrial Research Organisation, released forty-nine species of dung beetle, with the aim of reducing the amount of dung and therefore also the potential breeding sites of the fly.

Side-effects

Biological control can affect biodiversity through predation, parasitism, pathogenicity, competition, or other attacks on non-target species. An introduced control does not always target only the intended pest species; it can also target native species. In Hawaii during the 1940s parasitic wasps were introduced to control a lepidopteran pest and the wasps are still found there today. This may have a negative impact on the native ecosystem, however, host range and impacts need to be studied before declaring their impact on the environment.

Vertebrate animals tend to be generalist feeders, and seldom make good biological control agents; many of the classic cases of "biocontrol gone awry" involve vertebrates. For example, the cane toad (*Bufo marinus*) was intentionally introduced to Australia to control the greyback cane beetle (*Dermolepida albohirtum*), and other pests of sugar cane. 102 toads were obtained from Hawaii and bred in captivity to increase their numbers until they were released into the sugar cane fields of the tropic north in 1935. It was later discovered that the toads could not jump very high and so were unable to eat the cane beetles which stayed up on the upper stalks of the cane plants. However the toad thrived by feeding on other insects and it soon spread very rapidly; it took over native amphibian habitat and brought foreign disease to native toads and frogs, dramatically reducing their populations. Also when it is threatened or handled, the cane toad releases poison from parotoid glands on its shoulders; native Australian species such as goannas, tiger snakes, dingos and northern quolls that attempted to eat the toad were harmed or killed. However, there has been some recent evidence that native predators are adapting, both physiologically and through changing their behaviour, so in the long run, their populations may recover.

Rhinocyllus conicus, a seed-feeding weevil, was introduced to North America to control exotic musk thistle (*Carduus nutans*) and Canadian thistle (*Cirsium arvense*). However the weevil also attacks native thistles, harming such species as the endemic Platte thistle (*Cirsium neomexicanum*) by selecting larger plants (which reduced the gene pool), reducing seed production and ultimately threatening the species' survival.

The small Asian mongoose (*Herpestus javanicus*) was introduced to Hawaii in order to control the rat population. However it was diurnal and the rats emerged at night, and it preyed on the endemic birds of Hawaii, especially their eggs, more often than it ate the rats, and now both rats and mongooses threaten the birds. This introduction was undertaken without understanding the consequences of such an action. No regulations existed at the time, and more careful evaluation should prevent such releases now.

The sturdy and prolific eastern mosquitofish (*Gambusia holbrooki*) is a native of the southeastern United States and was introduced around the world in the 1930s and 40s to feed on mosquito larvae and thus combat malaria. However, it has thrived at the expense of local species, causing a decline of endemic fish and frogs through competition for food resources, as well as through eating their eggs and larvae. In Australia, the mosquitofish is the subject of discussion as to how best to control it; in 1989 it was said that "biological population control is well beyond present capabilities", and this remains the position.

Grower Education

A potential obstacle to the adoption of biological pest control measures is growers sticking to the familiar use of pesticides. It has been claimed that many of the pests that are controlled today using pesticides, actually became pests because pesticide use reduced or eliminated natural predators. A method of increasing grower adoption of biocontrol involves is letting growers learn by doing, for example showing them simple field experiments, having observations of live predation of pests, or collections of parasitised pests. In the Philippines, early season sprays against leaf folder caterpillars were common practice, but growers were asked to follow a 'rule of thumb' of not spraying against leaf folders for the first 30 days after transplanting; participation in this resulted in a reduction of insecticide use by 1/3 and a change in grower perception of insecticide use.

Mechanical Pest Control

Mechanical pest control is the management and control of pests using physical means such as fences, barriers or electronic wires. It includes also weeding and change of temperature to control pests. Many farmers at the moment are trying to find sustainable ways to remove pests without harming the ecosystem.

Methods

Handpicking

The use of human hands to remove harmful insects or other toxic material is often the most common action by gardeners. It is also classified as the most direct and the quickest way to remove clearly visible pests. However, it also has equal disadvantages as it must be performed before damage to the plant has been done and before the key development of insects.

Mechanical Traps

Mechanical traps or physical attractants are used in three main ways: to efficiently trap insects, to kill them or to estimate how much many insects there are in the total landmass using sampling method. However, some traps are expensive to produce and can end up benefiting insects rather than harming them.

Differences from Integrated Pest Control

Integrated pest control refers to the use of any means to control pests once they reach unacceptable levels. Mechanical pest control is but a minor part of integrated pest control. It means only the use of physical means to control pests.

References

- Nicole Davis (September 9, 2009). "Genome of Irish potato famine pathogen decoded". Haas et al. Broad Institute of MIT and Harvard. Retrieved 24 July 2012.

- "Scientists discover how deadly fungal microbes enter host cells". (VBI) at Virginia Tech affiliates. Physorg. July 22, 2010. Retrieved July 31, 2012.

- Jackson RW (editor). (2009). Plant Pathogenic Bacteria: Genomics and Molecular Biology. Caister Academic Press. ISBN 978-1-904455-37-0.

- "1st large-scale map of a plant's protein network addresses evolution, disease process". Dana-Farber Cancer Institute. July 29, 2011. Retrieved 24 July 2012.

- "Plasmopara viticola, the Cause of Downy Mildew of Grapes". The Origin of Plant Pathology and The Potato Famine, and Other Stories of Plant Diseases. Retrieved 4 February 2015.

- "Fusarium oxysporum : The End of the Banana Industry?". The Origin of Plant Pathology and The Potato Famine, and Other Stories of Plant Diseases. Retrieved 4 February 2015.

- "Help WildCare Pursue Stricter Rodenticide Controls in California". wildcarebayarea.org/. Wild Care. Retrieved 28 February 2014.

- "The Chinese Scientific Genius. Discoveries and inventions of an ancient civilization: Biological Pest Control" (PDF). The Courier. UNESCO: 24. October 1988. Retrieved 5 June 2016.

- Shapiro-Ilan, David I; Gaugler, Randy. "Biological Control. Nematodes (Rhabditida: Steinernematidae & Heterorhabditidae)". Cornell University. Retrieved 7 June 2016.

- "Conservation of Natural Enemies: Keeping Your "Livestock" Happy and Productive". University of Wisconsin. Retrieved 7 June 2016.

- Wilson, L. Ted; Pickett, Charles H.; Flaherty, Donald L.; Bates, Teresa A. "French prune trees: refuge for grape leafhopper parasite" (PDF). University of California Davis. Retrieved 7 June 2016.

- Acorn, John (2007). Ladybugs of Alberta: Finding the Spots and Connecting the Dots. University of Alberta. p. 15. ISBN 978-0-88864-381-0.

- Xu (2004). Combined Releases of Predators for Biological Control of Spider Mites Tetranychus urticae Koch and Western Flower Thrips Frankliniella occidentalis (Pergande). Cuvillier Verlag. p. 37. ISBN 978-3-86537-197-3.

- Capinera, John L. (October 2005). "Featured creatures:". University of Florida website - Department of Entomology and Nematology. University of Florida. Retrieved 7 June 2016.

Permissions

Index

www.ingramcontent.com/pod-product-compliance
Lightning Source LLC
Chambersburg PA
CBHW061313190326
41458CB00011B/3797

* 9 7 8 1 6 3 5 4 9 1 5 4 8 *